VEHICULAR 2020

The Ninth International Conference on Advances in Vehicular Systems, Technologies and Applications

October 18 – 22, 2020

VEHICULAR 2020 Editors

Akimasa Suzuki, Iwate Prefectural University, Japan
Yuping He, University of Ontario Institute of Technology, Canada
Parimala Thulasiraman, University of Manitoba, Canada

Printed from e-media with permission by:

Curran Associates, Inc.
57 Morehouse Lane
Red Hook, NY 12571

Some format issues inherent in the e-media version may also appear in this print version.

International Academy, Research, and Industry Association (IARIA)
412 Derby Way
Wilmington, DE 19810

Phone: (408) 893-6407
Fax: (408) 527-6351

petre@iaria.org

Additional copies of this publication are available from:

Curran Associates, Inc.
57 Morehouse Lane
Red Hook, NY 12571 USA
Phone: 845-758-0400
Fax: 845-758-2633
Email: curran@proceedings.com
Web: www.proceedings.com

Table of Contents

VEHICULAR 2020

Foreword

The Ninth International Conference on Advances in Vehicular Systems, Technologies and Applications (VEHICULAR 2020), held between October 18–22, 2020, continued the inaugural event considering the state-of-the-art technologies for information dissemination in vehicle-to-vehicle and vehicle-to-infrastructure and focusing on advances in vehicular systems, technologies and applications.

Mobility brought new dimensions to communication and networking systems, making possible new applications and services in vehicular systems. Wireless networking and communication between vehicles and with infrastructure have specific characteristics from other conventional wireless networking systems and applications (rapidly-changing topology, specific road direction of vehicle movements, etc.). These led to specific constraints and optimizations techniques; for example, power efficiency is not as important for vehicle communications as it is for traditional ad hoc networking. Additionally, vehicle applications demand strict communications performance requirements that are not present in conventional wireless networks. Services can range from time-critical safety services, traffic management, to infotainment and local advertising services. They are introducing critical and subliminal information. Subliminally delivered information, unobtrusive techniques for driver's state detection, and mitigation or regulation interfaces enlarge the spectrum of challenges in vehicular systems.

We take here the opportunity to warmly thank all the members of the VEHICULAR 2020 Technical Program Committee, as well as the numerous reviewers. The creation of such a high quality conference program would not have been possible without their involvement. We also kindly thank all the authors who dedicated much of their time and efforts to contribute to VEHICULAR 2020. We truly believe that, thanks to all these efforts, the final conference program consisted of top quality contributions.

Also, this event could not have been a reality without the support of many individuals, organizations, and sponsors. We are grateful to the members of the VEHICULAR 2020 organizing committee for their help in handling the logistics and for their work to make this professional meeting a success.

We hope that VEHICULAR 2020 was a successful international forum for the exchange of ideas and results between academia and industry and for the promotion of progress in the field of vehicular systems, technologies and applications.

VEHICULAR 2020 Chairs

VEHICULAR 2020 Steering Committee

Éric Renault, ESSIE Paris, France
Khalil El-Khatib, University of Ontario Institute of Technology - Oshawa, Canada

VEHICULAR 2020 Publicity Chair

Jose M. Jimenez, Universitat Politecnica de Valencia, Spain
Jose Luis García, Universitat Politecnica de Valencia, Spain

VEHICULAR 2020 Industry/Research Advisory Committee

Clément Zinoune, Renault, France
Yi Ding, US Army RDECOM-TARDEC, USA
William Whyte, Security Innovation, USA
Markus Ullmann, Federal Office for Information Security / University of Applied Sciences Bonn-Rhine-Sieg, Germany
Manabu Tsukada, University of Tokyo, Japan

VEHICULAR 2020

Committee

VEHICULAR 2020 Steering Committee

Éric Renault, Institut Mines-Télécom | Télécom SudParis, France
Khalil El-Khatib, University of Ontario Institute of Technology - Oshawa, Canada

VEHICULAR 2020 Publicity Chair

Jose M. Jimenez, Universitat Politecnica de Valencia, Spain
Jose Luis García, Universitat Politecnica de Valencia, Spain

VEHICULAR 2020 Industry/Research Advisory Committee

Clément Zinoune, Renault, France
Yi Ding, US Army RDECOM-TARDEC, USA
William Whyte, Security Innovation, USA
Markus Ullmann, Federal Office for Information Security / University of Applied Sciences Bonn-Rhine-Sieg, Germany
Manabu Tsukada, University of Tokyo, Japan

VEHICULAR 2020 Technical Program Committee

Nor Fadzilah Abdullah, National University of Malaysia (UKM), Malaysia
Mohammed Al-Ansi, University Malaysia Perlis (UniMAP), Malaysia
Sufyan T. Faraj Al-Janabi, University of Anbar, Ramadi,Iraq
Mustafa S. Al-Jumaily, University of Tennessee, Knoxville, USA
Ali Alfoudi, Al-Qadisiyah University, Irak
Ala'a Al-Momani, Ulm University, Germany
Bhaskar Anand, Indian Institute of Technology Hyderabad, India
Andrea Araldo, Telecom SudParis (Institut Polytechnique de Paris), France
Muhammad Asim, Chung-Ang University, Seoul, South Korea
Hakan Aydin, Karadeniz Technical University, Turkey
Eduard Babulak, Fort Hays State University, USA
Andrea Baiocchi, University of Roma "Sapienza", Italy
Ali Balador, RISE Research Institute of Sweden, Sweden
Paulo C. Bartolomeu, University of Aveiro, Portugal
Marcel Baunach, Graz University of Technology, Austria
Rahim (Ray) Benekohal, University of Illinois at Urbana-Champaign, USA
Sylvia Bhattacharya, Kennesaw State University, USA
Christos Bouras, University of Patras, Greece
Marcos F. Caetano, University of Brasilia, Brazil
Rodrigo Capobianco Guido, São Paulo State University (UNESP), Brazil

Florent Carlier, Centre de Reserche en Education de Nantes / Le Mans Université, France
Juan Carlos Ruiz, Universitat Politecnica de Valencia, Spain
Claude Chaudet, Webster University Geneva, Switzerland
Gihwan Cho, Jeonbuk University, Korea
Gianpiero Costantino, Institute of Informatics and Telematics (IIT) | National Research Council (CNR), Italy
Yousef-Awwad Daraghmi, Palestine Technical University-Kadoorie, Palestine
David de Andrés, Universitat Politècnica de València, Spain
Fawad Ud Din, McGill University, Canada
Liza Dixon, Independent Researcher, Germany
Suzi Iryanti Fadilah, Universiti Sains Malaysia, University
Mariano Falcitelli, Photonic Networks & Technologies National Laboratory of CNIT, Italy
Abraham O. Fapojuwo, University of Calgary, Canada
Gustavo Fernandez Dominguez, Center for Digital Safety & Security | AIT Austrian Institute of Technology, Austria
Miguel Franklin de Castro, Federal University of Ceará, Brazil
Tomonari Furukawa, University of Virginia, USA
Varun Garg, UMass Lowell, USA
Pedro Pablo Garrido Abenza, Universidad Miguel Hernandez de Elche, Spain
Apostolos Gkamas, University Ecclesiastical Academy of Vella of Ioannina, Greece
Sezer Goren, Yeditepe University, Turkey
Javier Gozalvez, Universidad Miguel Hernandez de Elche, Spain
Rami Hamdi, Hamad Bin Khalifa University, Qatar Foundation, Qatar
Kyungtae (KT) Han, Toyota North America, USA
Hong Hande, Huawei Technologies, Singapore
Yuping He, University of Ontario Institute of Technology, Canada
Javier Ibanez-Guzman, Renault S.A., France
Hocine Imine, IFSTTAR/LEPSIS, France
Uzair Javaid, National University of Singapore, Singapore
Dush Nalin Jayakody, Tomsk Polytechnic University, Russia
Terje Jensen, Telenor, Norway
Yiming Ji, Georgia Southern University, USA
Felipe Jiménez Alonso, Technical University of Madrid, Spain
Magnus Jonsson, Halmstad University, Sweden
Filbert Juwono, Curtin University, Malaysia
Gorkem Kar, Bahcesehir University, Turkey
Frank Kargl, Institute of Distributed Systems | Ulm University, Germany
Sokratis K. Katsikas, Norwegian University of Science and Technology, Norway
Norazlina Khamis, Universiti Malaysia Sabah, Malaysia
BaekGyu Kim, Toyota Motor North America Inc., USA
Zdzislaw Kowalczuk, Gdansk University of Technology, Poland
Ryo Kurachi, Nagoya University, Japan
Gyu Myoung Lee, Liverpool John Moores University, UK
Lianggui Liu, Zhejiang Shuren University, China
Tomasz Mach, Samsung R&D Institute, UK
Zoubir Mammeri, IRIT - Paul Sabatier University, France
P. Takis Mathiopoulos, University of Athens, Greece
Ioannis Mavromatis, University of Bristol, UK

Dalila B. Megherbi, University of Massachusetts, USA
Rashid Mehmood, King Abdul Aziz University,Saudi Arabia
Maria Luisa Merani, University of Modena and Reggio Emilia, Italy
Gerardo Mino Aguilar, Benemérita Universidad Autónoma De Puebla, Mexico
Naveen Mohan, KTH, Sweden
Bruno Monsuez, ENSTA ParisTech, France
Luís Moutinho, Escola Superior de Gestão e Tecnologia de Águeda (ESTGA) - University of Aveiro / Instituto de Telecomunicações, Portugal
Jose Eugenio Naranjo Hernandez, Universidad Politécnica de Madrid | Instituto Universitario de Investigación del Automóvil (INSIA), Spain
Ridha Nasri, Orange Labs, France
Keivan Navaie, Lancaster University, UK
Patrik Österberg, Mid Sweden University, Sweden
Antonio M. Pascoal, Institute for Systems and Robotics - IST | Univ. Lisbon, Portugal
Al-Sakib Khan Pathan, Southeast University, Bangladesh
Paulo Pinto, Universidade Nova de Lisboa, Portugal
Srinivas Pulugurtha, The University of North Carolina at Charlotte, USA
Hesham Rakha, Virginia Tech Transportation Institute, USA
Mohd Fadlee A Rasid, Universiti Putra Malaysia, Malaysia
Paulo Alexandre Regis, Southeastern Louisiana University, USA
Éric Renault, ESIEE Paris, France
Martin Ring, Bosch Engineering GmbH, Germany
Geraldo P. Rocha Filho, University of Brasília, Brazil
Justin P. Rohrer, Naval Postgraduate School, USA
Aymeric Rousseau, Argonne National Laboratory, USA
Javier Rubio-Loyola, CINVESTAV, Mexico
João Rufino, Instituto de Telecomunicações - Pólo Aveiro, Portugal
José Santa, Technical University of Cartagena, Spain
Nico Saputro, Parahyangan Catholic University, Bandung, Indonesia
Erwin Schoitsch, AIT Austrian Institute of Technology GmbH, Austria
Michele Segata, University of Trento, Italy
Alireza Shahrabi, Glasgow Caledonian University, Scotland, UK
Rajan Shankaran, Macquarie University, Australia
Prinkle Sharma, University of Massachusetts Dartmouth, USA
Dana Simian, Lucian Blaga University of Sibiu, Romania
Vasco N. G. J. Soares, Instituto de Telecomunicações / Instituto Politécnico de Castelo Branco, Portugal
Chokri Souani, Higher Institute of Applied Sciences and Technology of Sousse, Tunisia
Essam Sourour, Prince Sattam Bin Abdul-Aziz University (PSAU), Saudi Arabia
Mujdat Soyturk, Marmara University, Turkey
Anand Srivastava, IIIT Delhi, India
Qasim Sultan, Chung-Ang University, Seoul, South Korea
Akimasa Suzuki, Iwate Prefectural University, Japan
Daxin Tian, Beihang University, Beijing, China
Angelo Trotta, University of Bologna, Italy
Bugra Turan, Koc University, Istanbul, Turkey
Markus Ullmann, Federal Office for Information Security / University of Applied Sciences Bonn-Rhine-Sieg, Germany
KlausDavid, University ofKassel, Germany

Massimo Villari, Universita' di Messina, Italy
Ankur Vora, Intel, SanJose, USA
You-Chiun Wang, National Sun Yat-sen University, Taiwan
Ramin Yahyapour, Gesellschaft für wissenschaftliche Datenverarbeitung mbH Göttingen (GWDG), Germany
Shingchern D. You, National Taipei University of Technology, Taiwan
David Zage, Intel Corporation, USA
Sherali Zeadally, University of Kentucky, USA

Copyright Information

Study on Multi-Users Interference in Vehicle to Vehicle Visible Light Communications

Emmanuel Plascencia*+, Oyunchimeg Shagdar*, Hongyu Guan+ and Luc Chassagne+

*Institut VEDECOM, 23 bis Allée des Marronniers, 78000 Versailles, France
+Laboratoire d'Ingénierie des Systèmes de Versailles University of Versailles, IUT Velizy, France
Email:{luis-emmanuel.plascencia, oyunchimeg.shagdar}@vedecom.fr
Email:{hongyu.guan, luc.chassagne}@uvsq.fr

***Abstract*—Visible Light Communications (VLC) can play an important role in the Cooperative Intelligent Transport Systems (C-ITS) by enabling vehicles to communicate with nearby vehicles (V2V) and infrastructure (V2I) by offering virtually unlimited and unregulated spectrum. Whereas extensive R&D efforts have been made on physical layer techniques, almost no study has been made on Multi-Users Interference (MUI), consequently Medium Access Control (MAC), in VLC. This work sheds light on the impacts of MUI on VLC performances for V2V communications. We first develop an analytical model that formulates the Packet Delivery Ratio (PDR) performances of VLC communication in presence of MUI. We then conduct simulation evaluations to confirm the analytical model and evaluate the VLC performance when there is one or more interfering nodes. The obtained results clearly show that, in an absence of MAC, VLC can suffer from MUI in medium to dense traffic density even when message generation rate at each node is relately low.**

Keywords–Visible Light Communication (VLC); Multi-Users Interference (MUI), Vehicle to Vehicle (V2V) communication; mathematical modeling, MATLAB

I. Introduction

The key objective of C-ITS is to improve road safety and traffic efficiency by enabling vehicles and infrastructure to exchange information via Vehicle to Vehicle/Infrastructure communication (V2X). Radio communications technologies, particularly 802.11p and 4G/5G are considered to be the key players by providing omni-directional medium to long distance communications, allowing vehicles and roadside infrastructure communicate directly with each other (V2V or V2I) or through a network (V2N). Whereas radio communication technologies are probably the *de-facto* choice for a great number of C-ITS applications, due to the limited radio resource and their vulnerability against security attacks, there is a need for complementary technology especially for applications that require 100% of reliability and strong cyber-security protection. Indeed applications particularly those for automated driving have extremely strict requirements in terms of reliability and security. One of such applications having such stringent requirements is vehicle platooning in which, V2V communication between platoon members are required for longitudinal and lateral controls avoiding chain instability problem. Because the information is to control vehicles, the information exchange has to be extremely reliable and secure. Targeting applications, such as vehicles platooning, a great number of researchers suggest VLC for V2V as a complementary solution to radio communications as shown in Figure 1, [1]–[5]. VLC uses the visible light spectrum (wavelengths between 780 nm to 375 nm) as communication media. VLC is a fast, safe and cheap technology, since it is implemented directly using vehicle headlights and taillights. Moreover, the usage of Light Emitter Diodes (LED) instead of a xenon or halogen bulb, has a number of benefits such as long useful life, low power consumption, high tolerance to environmental conditions, and high efficiency [6]. Finally, since radio and light communications do not interfere with each other, and hence VLC can perfectly co-exist and complement radio communications.

Indeed, IEEE provides the possibility to deploy VLC by specifying VLC standards: 802.15.7 [7] and 802.15.7r1 [8], which are published in 2011 and in 2018, respectively, and the ongoing work on IEEE 802.11bb [9]. The standard IEEE 802.15.7 and its revision 802.15.7r1 have a strong focus physical layer configurations of VLC for both indoor and outdoor LED to photodiode (PD) or LED to camera VLC communications. Concerning the MAC, the standards basically carried over the solutions of Wireless Personal Area Networks (WPAN). On the other hand, the aim of IEEE 802.11bb [9] is to integrate Wireless Local Area Network (WLAN) solutions to Light Fidelity communication (LiFi). In adding to the above-mentioned standardisation efforts, a great number of R&D studies have been carried out proposing VLC as a candidate technology for enhanced reliability and security for V2X communications. The majority of the efforts however focuses on the physical layer design of VLC proposing modulation schemes, filtering strategies, etc. [10]–[12]. Some real-world demonstrations of VLC prototypes for V2V communication have been also made [1], [2], [12]. In contrast to the voluminous literature on physical layer solutions, very few efforts

Figure 1. Visible Light Communication for V2V information exchange.

are made on medium access control for VLC [13]–[16]. Most importantly, because VLC is directional and requiring Line Of Sight (LOS) condition, one may even doubt about MUI in VLC, and hence neglecting the importance of MAC.

To the best of our knowledge, the current work is the only work that studies the impact of MUI in VLC. In this paper, we first develop analytical models that formulate the size of MUI zone in V2V VLC. We then further develop a model of PDR in VLC in presence of MUI, when vehicles' density follows the Poisson distribution. Finally, by using computer simulations, we validate the correctness of the theoretical model and evaluate the VLC performance in presence of MUI. The results clearly show that PDR can quickly degrade down to few % for medium to dense roads.

This paper is organized as follows. Section II highlights the related work. We develop an analytical model on impacts of MUI on V2V VLC in Section III. Section IV validates the analytical model and evaluates the PDR performance of VLC up to 3- and 7-lanes of highway scenarios with sparse to dense traffic density. Finally, Section V concludes the paper.

II. Related Work

The majority of the existing standardisation and R&D efforts on VLC are to advance physical layer design of VLC. IEEE 802.15.7 standard specified three PHY modes PHY-I, II, and III, where PHY-I is intended to outdoor applications utilizing On-Off Keying (OOK) and Variable Pulse Position Modulation (VPPM) coding schemes, which are relatively robust in harsh outdoor environments. The authors of [10], [11] demonstrated different modulation schemes for transmission (Manchester, Orthogonal Frequency Division Multiplexing (OFDM), Miller, etc.), and complex filtering strategies for reception. *A. Belle et al.* [17] presented a VLC prototype of IEEE 807.15.7 that extending their previous work [18] on IEEE 802.15.4 [19]. The authors developed software libraries for PHY and *zigbee* MAC, which is not specifically designed for VLC. *Q. Wang et al.* [20] evaluated VLC performances, using a VLC platform, in terms of communication speed, communication distance, and low power LED/PD saturation in outdoor/indoor environments for different types of VLC systems: high/low power LED to PD or LED to LED. The authors of [1], proposed to use visible light not only for V2V communications but also for inter-vehicle distance measurement. Finally, a number of demonstrations using VLC for V2V communications have been made [1], [2].

In contrast to the great number of efforts made on PHY, very few works on MAC can be found in the literature. The authors of [13] studied Carrier Sense Multiple Access with Collision Avoidance (CSMA/CA) on top of the physical layer of specified by IEEE 802.15.7 [7] with different priority levels setting (High, Medium, and Low Priorities), which result in differentiated settings of back-off time, back-off exponent and contention window sizes. *P. Shams, et al.* [15] presented a performance evaluation of throughput, delay, power consumption, collision probability, transmission probability, access probability and packet discard probability based on Markov modelling and MATLAB simulations of IEEE 802.15.7 VLC standard. The authors of [21] evaluated the service time distribution of the IEEE 802.15.7 standard using Markov chain model. The authors also proposed an analytic and semi-analytic approach of queue modeling. *S. Ishihara et al.* [14] proposed a

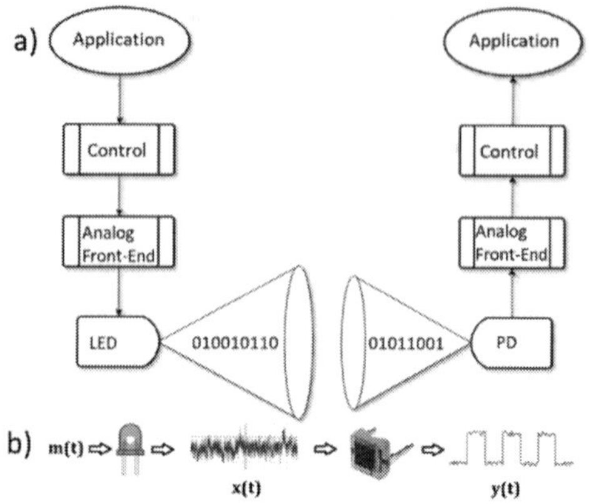

Figure 2. VLC System, a) VLC Schematic b) Conceptualization of the VLC channel.

radio and visible light hybrid communication for platooning applications implementing IEEE 1609.4/802.11p standard for radio frequency and ALOHA MAC protocol for VLC. None of the above efforts are based on studies on MUI in VLC, demonstrating the needs of MAC. Furthermore, the rational behind the usages of CSMA/CA or ALOHA for VLC, which is naturally half-duplex and directional communication, is not clear. This paper is to fill the missing gap, studying MUI in VLC, presenting the need of MAC, which shall take into account the VLC properties.

Concerning VLC evaluation tools, *Q. Wang et al.* [20] presented a low cost, flexible and open source VLC platform enabling researchers to develop and test their own VLC systems. Besides prototyping, researchers can also conduct simulations. Veins VLC [22], [23] is a simulation framework that integrates VLC transmitter, receiver, and channel models. Nevertheless, it misses several important functionalities including that transmission power ($10mW$) cannot be reconfigured, and the noise takes into account only the thermal noise ignoring environmental noises. For this reasons, in this paper, we have developed own VLC models using Simulink tool.

To summarize, motivated by the fact that there is no work studying MUI in VLC, this paper is dedicated to study on MUI and presents the need of MAC that takes into account the VLC properties.

III. Modeling Impact of VLC Multi-users Interference

In this Section, we first present a VLC channel model and then develop an analytical model of a zone, from which no interference is allowed such that an ongoing VLC communication is protected. Note that for the sake of simplicity, the model does not take into account weather condition, sun light direction and intensity.

A. VLC Channel Model

Intensity Modulation with Direct Detection (IM/DD) is a commonly used method for optical communications. As illustrated in Figure 2, the LED emits the modulated signal $m(t)$, whose intensity is varied in accordance with the data. After propagating through the wireless channel, the signal $x(t)$ is collected by the photo-diode (PD) of the receiver. The latter generates a current, $y(t)$, which is proportional to the power of the light incident on the active area of the PD.

In order to model such a VLC [1], [2], [24], we first need to express the angular distribution ($R_o(\phi)$) or intensity pattern generated by the LED:

$$R_o(\phi) = \begin{cases} \frac{(m_i+1)}{2\pi} \cos^{m_i}(\phi) & \phi \in [\frac{-\pi}{2} \frac{\pi}{2}]. \\ 0 & \phi \geq \frac{\pi}{2}, \end{cases} \quad (1)$$

Here, m_i is the Lambert coefficient related to the LED semi-angle at half-power $\phi_{\frac{1}{2}}$ (see Figure 3):

$$m_i = \frac{-\ln 2}{\ln(\cos \phi_{\frac{1}{2}})} \quad . \quad (2)$$

For the receiver side, its effective reception area ($A_{eff}(\psi)$) is modeled:

$$A_{eff}(\psi) = \begin{cases} A_r \cos(\psi) & 0 \leq \psi \leq \frac{\pi}{2}, \\ 0 & \psi > \frac{\pi}{2} \end{cases} \quad (3)$$

where, A_r is the active area, collecting the light beams at angles ψ (see Figure 3).

The wireless optical channel then, can be modeled considering the wireless link with an array of several LEDs without optical lenses as the transmitter and a PD as the receiver. The

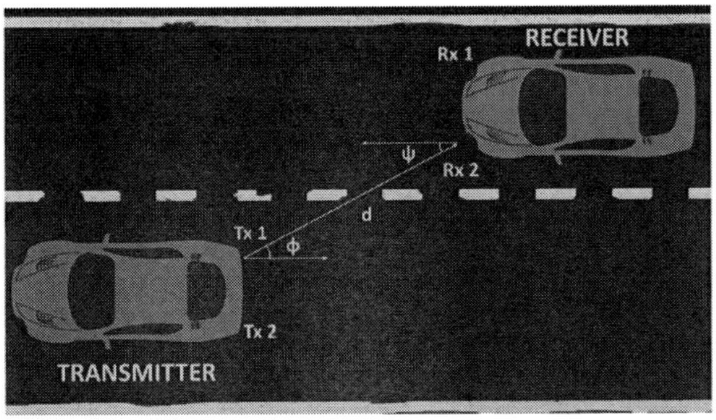

Figure 3. Relative positioning of a transmitter (blue vehicle) and a receiver (red vehicle).

DC gain ($H(\phi, \psi)$) is expressed as follows for a PD placed at d distance with an irradiance (ϕ) and an incidence (ψ) angles [1], [2], [24] (see Figure 3):

$$H(\phi, \psi) = R_o(\phi) \cdot Aeff(\psi) 0 \leq \psi \leq \psi_c \quad , \quad (4)$$

$$H(\phi, \psi) = \begin{cases} \frac{A_r(m_i+1)}{2\pi d^2} \cos^{m_i}(\phi) \cos \psi & 0 \leq \psi \leq \psi_c, \\ 0 & \text{elsewhere} \end{cases} \quad . \quad (5)$$

The receiver power P_r, is hence

$$P_r = H(\phi, \psi) P_t = \frac{H_0(\phi, \psi)}{d^2} \quad , \quad (6)$$

where, P_t is the transmission power and $H_0(\phi, \psi)$ is

$$H_0(\phi, \psi) = \begin{cases} \frac{A_r(m_i+1)P_t}{2\pi} \cos^{m_i}(\phi) \cos \psi & 0 \leq \psi \leq \psi_c, \\ 0 & \text{elsewhere} \end{cases} \quad . \quad (7)$$

For the sake of simplicity, we assume that no error-correction coding applied. In such a case, as one can neglect the multipath fading in VLC [25], the ability of correctly decoding the received signal at the receiver, PDR, depends on the Bit-Error Rate (BER) and the packet size, L bits [26]:

$$PDR = (1 - BER)^L \quad . \quad (8)$$

The relation between BER and the Signal to Interference Noise Ratio (SINR) for OOK is expressed as follows [24]

$$BER = Q(\sqrt{SINR}) = Q(\sqrt{\frac{P_r}{MUI + N}}) \quad , \quad (9)$$

where Q function is defined as

$$Q(z) = \int_z^{\infty} \frac{1}{\sqrt{2\pi}} e^{-\frac{y^2}{2}} dy \quad . \quad (10)$$

In (9), N is the noise power, which consists of shot noise and thermal noise, and MUI is the total interference power. It is obvious that in order to correctly receive the transmitted data, the received SINR must be higher than a given threshold ($SINR_{th}$), which is determined by the modulation scheme.

$$\frac{P_r}{MUI + N} \geq SINR_{th} \quad . \quad (11)$$

B. *Impact of MUI*

For a given pair of intended transmitter and receiver, we are now interested in determining the geographical zone from where no interference is allowed, i.e., MUI zone. In another words, we will calculate the distance from an interfering node to the intended receiver (d_{ir}) that fulfills the following condition

$$P_i(d_{ir}) \geq \frac{P_r(d_{tr})}{SINR_{th}} - N \quad , \quad (12)$$

where d_{tr} is the distance between the intended transmitter and the receiver. P_i is the interference power, i.e., the receive power from the interfering node. Since VLC is directional, the transmitter and the interfering node (LEDs) have to be in the Field of View (FoV) of the receiver (PD), respecting the conditions of irradiance and incidence angles (see Figure 3).

As (12) suggests, we now need to determine the SINR threshold ($SINR_{th}$), which depends on the desired communication quality i.e., the PDR requirement. Since PDR=1 - PER, we can easily calculate the SINR threshold, using the equations (8) and (9) for a binary modulation scheme:

$$SINR_{th} = \left(Q^{-1}(1 - \sqrt[L]{PDR_{req}})\right)^2 \quad . \quad (13)$$

Here, Q^{-1} is the inverse Q function. Calculating (13) for the packet size of 1000 Bits, we can draw Figure 4, which shows $SINR_{th}$ (in dB) for different PDR requirements. As can be seen in the Figure, the SINR threshold sharply increases with the increase of the PDR threshold, taking 12.47 dB for the PDR requirement of 90%. We are now ready to determine the MUI zone for communication between an intended transmitter and a receiver. Because $\frac{P_r}{SINR_{th}} \gg N$ (indeed, we can observe $\frac{P_r}{SINR_{th}} = 1\mu W$ while $N = 50nW$), we can safely ignore N in (12).. Furthermore, since P_r and P_i are both expressed by

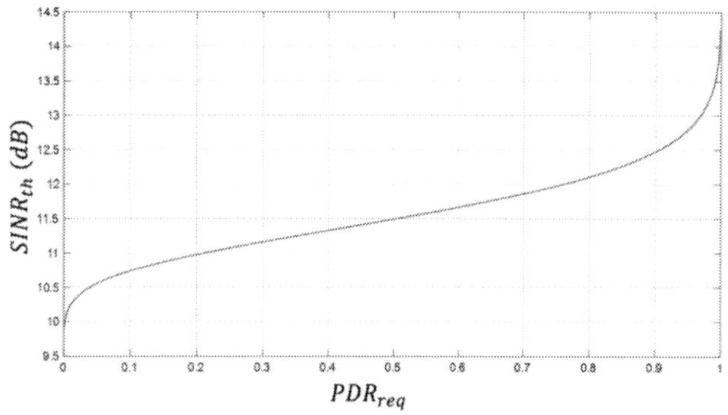

Figure 4. $SINR_{th}(dB)$ vs. PDR_{req}. Mathematical computation of the SINR threshold as a function of the PDR requirement. Here L=1000 Bits.

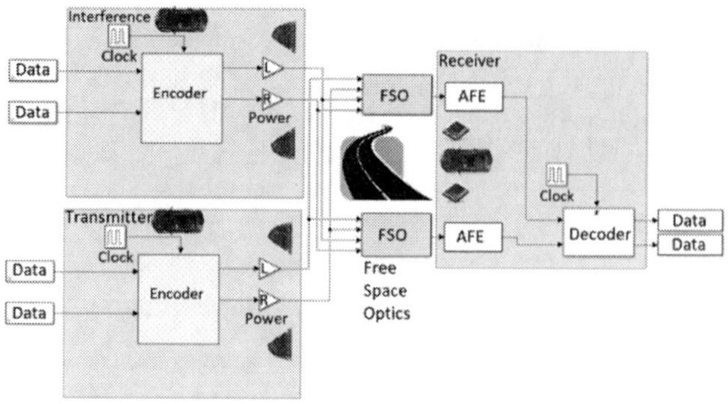

Figure 6. Simulink models for simulation evaluations.

(6) using d_{tr} and d_{ir}, respectively, the maximum d_{ir} satisfying the condition (12) is found as

$$d_{ir} = d_{tr}\sqrt{SINR_{th}} \quad . \tag{14}$$

Since the interfering nodes (as well as the intended transmitter) have to be in the FoV of the receiver, for a given pair of an intended transmitter (Tx) and receiver (Rx), the MUI zone is the circular sector with the radius d_{ir} and the central angle 2ψ as shown in orange in Figure 5. If we consider V2V

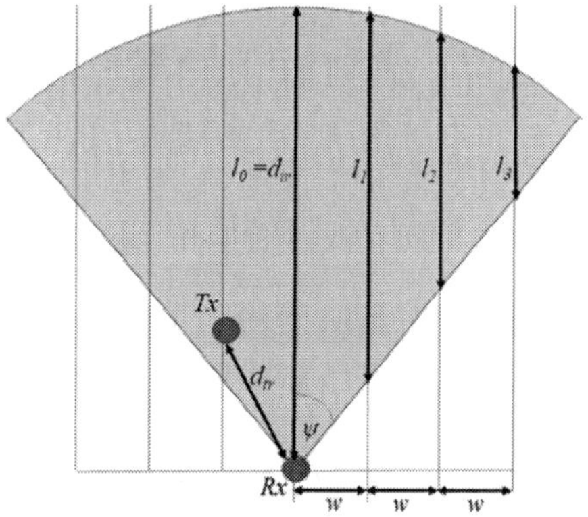

Figure 5. The MUI zone for a given pair of transmitter (Tx) and receiver of VLC (Rx).

communication on multi-lane straight road, vehicles can be only on individual lanes. For such a scenario, we now calculate the lengths of each lane belonging to the MUI zone (see Figure 5). As d_{ir} is the distance from the receiver to the limit of the MUI zone (see (14)), we can assume $d_{ir} \gg w$, where w is the lane width, the lengths of individual lanes l_k in the MUI zone (see Figure 5) are

$$l_k = d_{ir} - k \cdot w \cdot \cot\psi \quad , \quad 0 \leq k \leq n \quad , \tag{15}$$

where n is the number of lanes on the right or left side of the receiver. The total length of lanes in the MUI zone is then

$$l = d_{ir} + \sum_{k=1}^{n_l} l_k + \sum_{k=1}^{n_r} l_k \quad . \tag{16}$$

Here n_l (resp. n_r) is the number of adjacent lanes on left (resp. right) side of the receiver.

Targeting an intended transmitter and a receiver are at d_{tr} distance with the LOS condition, we now calculate the probability of having interfering vehicles on the lanes in the MUI zone. More specifically, we assume that vehicles on a highway follow Poisson distribution, which is widely used for highway traffic [27], the probability of finding i vehicles on the l length of the road is [28]

$$P(i,l) = \frac{(\beta l)^i e^{-\beta l}}{i!} \quad . \tag{17}$$

The communication between the intended transmitter and receiver will be successful only if none of the interfering vehicles in the MUI zone accesses the channel at the same time, i.e.,

$$P_s = \sum_{i=0}^{\infty} P(i,l)(1-\tau)^i \quad . \tag{18}$$

Here τ is the channel access probability at individual vehicles. Assuming that vehicles periodically generate messages (beacons) and letting T_{tx} be the average packet transmission time, τ is found as

$$\tau = \frac{T_{tx}}{T_{interval}} \quad . \tag{19}$$

Here $T_{interval}$ is the message generation interval. It should be noted that, since we ignore shadowing induced by vehicles, (18) is the minimum success probability.

IV. PERFORMANCE EVALUATION

In this Section, we evaluate the impact of MUI on VLC using both the theoretical and simulation evaluations. For simulation evaluations, we have developed the models of VLC transmitter, receiver, and interfering nodes using Matlab/SimuLink. As illustrated in Figure 6, the transmitter and interfering nodes have a message generator, a Digital to Analog Converter (DAC), and LEDs. They periodically generate sequences of bits, encode them using the Manchester modulation, and emit the signal using the LEDs. The receiver node has a PD and an Analog to Digital Converter (ADC), and hence tries to decode the received VLC signals. The simulation parameters are listed in Table I that are typical values of VLC based on LEDs to PDs. We first evaluate PDR for VLC communications from the intended transmitter to the receiver in absence of interfering nodes. Specifically, the simulations are carried out targeting on a 7-lane straight road, where the receiver (Rx) is

TABLE I. SIMULATION PARAMETERS.

Parameter	*Value*
PD reference	$S6967$ Hamamatsu [29]
A_{eff}	$100mm$ x $100mm$
PD efficiency	$0.5(A/W)$
FoV (ψ)	$55°$
PD capacitance	$1.12\mu F/m^2$
Transmission frequency	$500kHz$
Transmission power	1 Watt (car taillight)
Transmitter Semi-angle ($\phi_{1/2}$)	$20°$
Inter-PD separation distance	$1.2meters$
Road lane width (w)	$2.5meters$
Data size (L)	$1000Bytes$

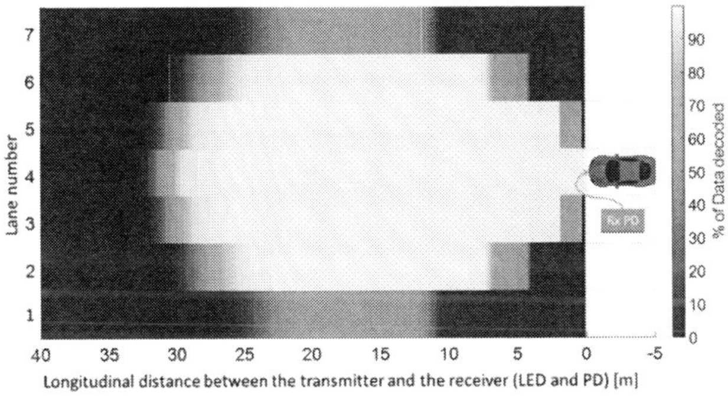

Figure 7. Communication range (PDR) of vehicle to vehicle VLC when transmitter and receiver are on a 7-lanes road.

fixed on the central lane (lane 4) and the intended transmitter (Tx) takes different positions. Figure 7 shows the obtained results, where horizontal and vertical axis are the longitudinal (i.e., longitudinal distance from the receiver) and the lateral (i.e., lane number) positions of Tx. The obtained PDR result for each position of Tx is expressed by the color plate. As

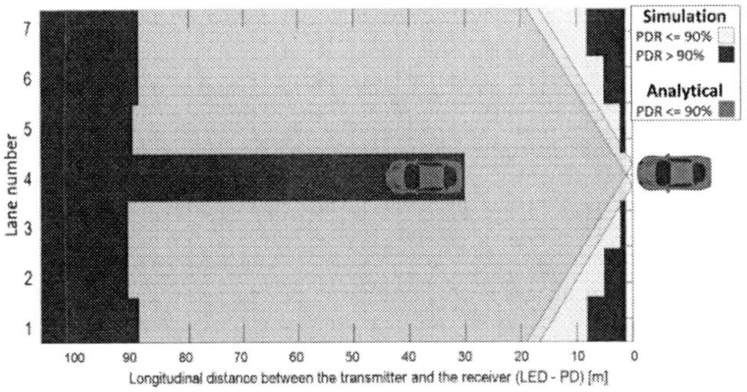

Figure 8. MUI zone for 90% of PDR requirement. Blue and yellow zones are simulation results, Red transparent area is the results of the analytical model.

expected, the longest communication distance is obtained when the transmitter is on the same lane as the receiver (i.e., the lateral distance is 0 meters). Specifically, 90% of PDR can be obtained up to 30 meters of longitudinal distance. On the other hand, when the transmitter is on the right or the left adjacent lanes, 90% of PDR can be obtained for 2 to 28 meters of longitudinal distances. When the transmitter is on a second adjacent lane, 90% of PDR can be obtained for

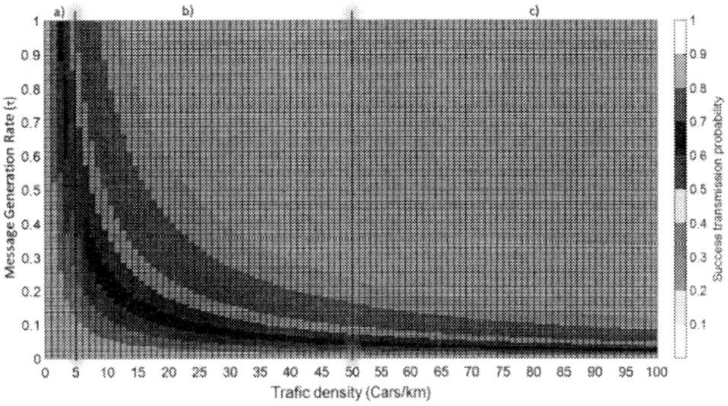

Figure 9. PDR performances of VLC in 3-lanes highway scenario ($l = 203m$). a) Low Density - maximum 1 vehicles on the road, b) Medium Density - maximum 10 vehicles on the road, and c) High Density - maximum 20 vehicles on the road.

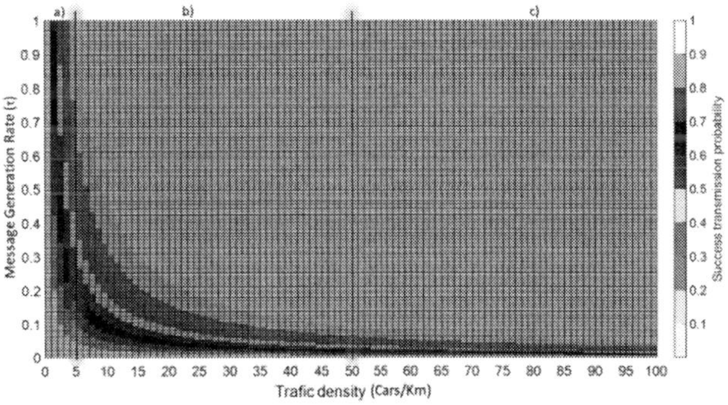

Figure 10. PDR performances of VLC in 7-lanes highway scenario ($l = 529m$). a) Low Density - maximum 2 vehicles on the road, b) Medium Density - maximum 26 vehicles on the road, and c) High Density - maximum 53 vehicles on the road.

the longitudinal distances between 7 to 26 meters. Finally, 90% of PDR cannot be obtained if the transmitter is a third adjacent lane. We now evaluate the impact of interference for the VLC communication. We fix the intended transmitter on the same lane as the receiver at 29 meters and evaluate PDR for varying positions of the interfering node. The obtained results are shown in Figure 8, in which two areas are depicted: yellow and blue. When the interfering node is in the yellow area, the PDR performance higher than 90% is not achieved. In contrast, when the interfering node is in the blue area, the communication between the intended Tx and Rx is succeeded having higher than 90% of PDR. The Figure also shows the analytical result by the red transparent area, which the MUI zone corresponds to PDR_{req} = 90% (see 13). As the figure shows, the simulation and analytical results match pretty well. Nevertheless, 10 meters of shift in the longitudinal direction is observed for the analytical results w.r.t that of the simulations. This is probably because a bit wider the FoV angle is used for the simulations. We now evaluate PDR results between the intended transmitter and the receiver, targeting the possibility of having more than one interfering vehicle on the road, and the vehicles density follows the Poisson distribution. Figures 9 and 10 show the results for 3- and 7-lanes highway scenarios, respectively. The horizontal axis of the figures is the traffic density, β in (17), and the vertical axis is the message generation rate, τ in (18). Finally, PDR results are depicted using the color plate. The results clearly show that, if there is no MUI control, i.e., no MAC, sufficient PDR performances are achievable only for extremely low traffic density and very low message generation rate. Particularly, in order to obtain 90% of PDR for the message generation rate of 0.9, the traffic density has to be lower than 1 and 2 cars/Km for 3-lanes and 7-lanes scenarios respectively. Otherwise, in order to obtain higher than 90% PDR in 50 cars/km density, the message generation rate has to be lower than 0.05 and 0.01 for 3- and 7-lanes scenarios, respectively. Although the obtained results are rather pessimistic figures, they clearly present a strong need of MAC protocol for VLC.

V. Conclusion

In this paper, we have developed an analytical model determining MUI zone, from which no interference is allowed for protecting an ongoing visible light V2V communication. Furthermore, we analytically formulated the PDR performance of a VLC when the interfering nodes follow the Poisson distribution. Finally, we have conducted simulation evaluations using a Simulink model. The model integrates, in adding to an intended transmitter, one or several interference vehicles, all of them with the possibility to generate, encode, and send random messages over the same VLC channel to the target receiver. Using the channel model for Free Space Optical (FSO) communications, and integrating the physical and electrical properties of a typically used PD, we have compared the results obtained by simulation and the developed analytical model. The simulation results first confirm the correctness of the analytical model on MUI zone. The results further show that even with low traffic densities and low message generation rates, the vehicles in the MUI zone can significantly degrade the PDR performance of the target VLC communication and presenting the inherent necessity of a MAC protocol for V2V communications. The future work includes an improvement of the theoretical model by taking into account shadowing effect by bodies of vehicles. We also conduct study on MAC that is aware of the presences of vehicles in the MUI zones by e.g., introducing two-hop beacons.

References

[1] B. Béchadergue, "Visible light range-finding and communication using the automotive led lighting," Ph.D. dissertation, 2017.

[2] M. Abualhoul, "Visible light and radio communication for cooperative autonomous driving: applied to vehicle convoy," Ph.D. dissertation, MINES ParisTech, 2016.

[3] J.-H. Yoo et al., "Demonstration of vehicular visible light communication based on led headlamp," International journal of automotive technology, vol. 17, no. 2, 2016, pp. 347–352.

[4] H.-Y. Tseng et al., "Characterizing link asymmetry in vehicle-to-vehicle visible light communications," in 2015 IEEE Vehicular Networking Conference (VNC). IEEE, 2015, pp. 88–95.

[5] M. D. Hina, H. Guan, A. Ramdane-Cherif, and N. Deng, "Secured data processing, notification and transmission in a human-vehicle interaction system," in 2016 IEEE 19th International Conference on Intelligent Transportation Systems (ITSC). IEEE, 2016, pp. 1277–1284.

[6] R. D. Dupuis and M. R. Krames, "History, development, and applications of high-brightness visible light-emitting diodes," Journal of lightwave technology, vol. 26, no. 9, 2008, pp. 1154–1171.

[7] 802.15.7-2011 - IEEE Standard for Local and Metropolitan Area Networks–Part 15.7: Short-Range Wireless Optical Communication Using Visible Light, IEEE Std., Sep. 2011, [Retrieved: May, 2020]. [Online]. Available: https://standards.ieee.org/standard/802_15_7-2011.html

[8] 802.15.7-2018 - IEEE Standard for Local and metropolitan area networks–Part 15.7: Short-Range Optical Wireless Communications, IEEE Std., Apr. 2019, [Retrieved: May, 2020]. [Online]. Available: https://standards.ieee.org/standard/802_15_7-2018.html

[9] 802.11 Light Communications Amendment - Task Group "bb", IEEE Std., [Retrieved: May, 2020]. [Online]. Available: http://www.ieee802.org/11/Reports/tgbb_update.htm

[10] M. Uysal, F. Miramirkhani, O. Narmanlioglu, T. Baykas, and E. Panayirci, "Ieee 802.15. 7r1 reference channel models for visible light communications," IEEE Communications Magazine, vol. 55, no. 1, 2017, pp. 212–217.

[11] A. Galisteo, D. Juara, and D. Giustiniano, "Research in visible light communication systems with openvlc1. 3," in 2019 IEEE 5th World Forum on Internet of Things (WF-IoT). IEEE, 2019, pp. 539–544.

[12] S.-A. Avătămăniţei et al., "Intensive testing of infrastructure-to-vehicle visible light communications in real outdoor scenario: Evaluation of a 50 meters link in direct sun exposure," in 2019 Global LIFI Congress (GLC). IEEE, 2019, pp. 1–5.

[13] V. Van Huynh, Y. M. Jang et al., "Priority mac based on multi-parameter for ieee 802.15. 7 vlc," in ICTC 2011. IEEE, 2011, pp. 257–260.

[14] S. Ishihara, R. V. Rabsatt, and M. Gerla, "Improving reliability of platooning control messages using radio and visible light hybrid communication," in 2015 IEEE Vehicular Networking Conference (VNC). IEEE, 2015, pp. 96–103.

[15] P. Shams, M. Erol-Kantarci, and M. Uysal, "Mac layer performance of the ieee 802.15. 7 visible light communication standard," Transactions on Emerging Telecommunications Technologies, vol. 27, no. 5, 2016, pp. 662–674.

[16] Q. Mao, P. Yue, M. Xu, Y. Ji, and Z. Cui, "Octmac: A vlc based mac protocol combining optical cdma with tdma for vanets," in 2017 International Conference on Computer, Information and Telecommunication Systems (CITS). IEEE, 2017, pp. 234–238.

[17] A. Bellè, M. Falcitelli, M. Petracca, and P. Pagano, "Development of ieee802. 15.7 based its services using low cost embedded systems," in 2013 13th International Conference on ITS Telecommunications (ITST). IEEE, 2013, pp. 419–425.

[18] R. Corsini et al., "Free space optical communication in the visible bandwidth for v2v safety critical protocols," in 2012 8th International Wireless Communications and Mobile Computing Conference (IWCMC). IEEE, 2012, pp. 1097–1102.

[19] 802.15.4-2015 - IEEE Standard for Low-Rate Wireless Networks, IEEE Std., Apr. 2016, [Retrieved: May, 2020]. [Online]. Available: https://standards.ieee.org/standard/802_15_4-2015.html

[20] Q. Wang, D. Giustiniano, and O. Gnawali, "Low-cost, flexible and open platform for visible light communication networks," in Proceedings of the 2nd International Workshop on Hot Topics in Wireless, 2015, pp. 31–35.

[21] K. A. Mehr, S. K. Nobar, and J. M. Niya, "Ieee 802.15. 7 mac under unsaturated traffic: Performance analysis and queue modeling," Journal of Optical Communications and Networking, vol. 7, no. 9, 2015, pp. 875–884.

[22] A. Memedi, H.-M. Tsai, and F. Dressler, "Impact of realistic light radiation pattern on vehicular visible light communication," in GLOBECOM 2017-2017 IEEE Global Communications Conference. IEEE, 2017, pp. 1–6.

[23] A. Memedi, C. Tebruegge, J. Jahneke, and F. Dressler, "Impact of vehicle type and headlight characteristics on vehicular vlc performance," in 2018 IEEE Vehicular Networking Conference (VNC). IEEE, 2018, pp. 1–8.

[24] Z. Ghassemlooy, W. Popoola, and S. Rajbhandari, Optical wireless communications: system and channel modelling with Matlab®. CRC press, 2019.

[25] M. Kavehrad and R. Aminikashani, Visible Light Communication Based Indoor Localization. CRC Press, 2019.

[26] J. Zhang, L. Cheng, and I. Marsic, "Models for non-intrusive estimation of wireless link bandwidth," in IFIP International Conference on Personal Wireless Communications. Springer, 2003, pp. 334–348.

[27] R. J. Salter, Highway traffic analysis and design. Macmillan International Higher Education, 1996.

[28] O. Shagdar, F. Nashashibi, and S. Tohme, "Performance study of cam over ieee 802.11 p for cooperative adaptive cruise control," in 2017 Wireless Days. IEEE, 2017, pp. 70–76.

[29] Plastic Si PIN Photodiode, Hamamatsu, 11 2014, [Retrieved: May, 2020]. [Online]. Available: https://www.hamamatsu.com/resources/pdf/ssd/s2506-02_etc_kpin1048e.pdf

Simulation Evaluation of Cooperative Intersection Traversing Method for Connected Vehicles

Koki Higashiyama, Kenta Kimura, Habibullah Babakarkhail, and Kenya Sato

Computer and Information Science
Graduate School of Science and Engineering
Doshisha University
Kyoto, Japan
Email: koki.higashiyama@nislab.doshisha.ac.jp, kenta.kimura@nislab.doshisha.ac.jp,
habibullah.babak@nislab.doshisha.ac.jp, ksato@mail.doshisha.ac.jp

***Abstract*—Connected vehicles can exchange information with each other by communicating via a network, enabling them to detect vehicles in their blind spots which cannot be seen by a vehicle alone, thereby contributing to traffic efficiency and safety. Since it will take time for such vehicles to become prevalent on roads, connected and non-connected vehicles will share the road in the future. We have developed a method that enables connected vehicles to share information gathered by their sensors on surrounding vehicles near an intersection. Simulation experiments were used to consider safety and evaluate changes in efficiency as the connected vehicle penetration rate increased. We found that safety can be ensured by adjusting the Time-To-Collision parameter dynamically, and that efficiency for an intersection with average traffic volume was improved compared with using conventional methods.**

Keywords–connected vehicle; cooperative automated driving; V2V communication; mixed traffic.

I. INTRODUCTION

Connected vehicles can exchange information with surrounding vehicles and roadside infrastructures using communication methods. Examples of these communication methods include Dedicated Short Range Communications (DSRC) and Cellular Vehicle-to-Everything (C-V2X) communication. DSRC is already being used in toll collection systems on expressways and in services that provide traffic information while C-V2X is being evaluated for practical use [1]. As well as such Vehicle-to-Infrastructure (V2I) communication, these communication technologies will be used for Vehicle-to-Vehicle (V2V) communication. Auto manufacturers are going to produce vehicles featuring V2V communication services for advanced safe driving support [2]. V2V communication enables connected vehicles to sense situations that cannot be recognized from only the vehicle's sensor information. A cooperative Intelligent Transport System (ITS) achieved through a combination of connected vehicles and autonomous driving technology, will enable traffic to flow more efficiently and safely.

Since it will take time for connected vehicles to become prevalent on roads, we can expect connected and non-connected (conventional) vehicles, which cannot communicate with other vehicles, to share the same roads. In the environment with only connected vehicles, the driving information (position, speed, etc.) of all vehicles on the road can be shared, thereby each vehicle is able to know where the other vehicles are traveling and which way they will go. That way, for example, at an intersection, each vehicle can detect in advance the presence of other vehicles approaching the intersection from the intersecting road. Then, if necessary, stop before the intersection to avoid a collision or, if there is no approaching vehicle, the vehicle can pass without stopping. Therefore, the safety and efficiency can be easily improved. On the other hand, in the mixed situation, while it is possible to share the information between connected vehicles, it is not possible to obtain the information of non-connected vehicles. Connected vehicles cannot know where non-connected vehicles are traveling. That is why, if the approaching vehicles at the intersection are non-connected vehicles, the connected vehicle cannot be able to detect the presence of them, and the improvement in safety and efficiency is incomplete. Therefore, it is necessary to develop methods that enables connected vehicles to share and use not only the information of each other but also that of the non-connected vehicles.

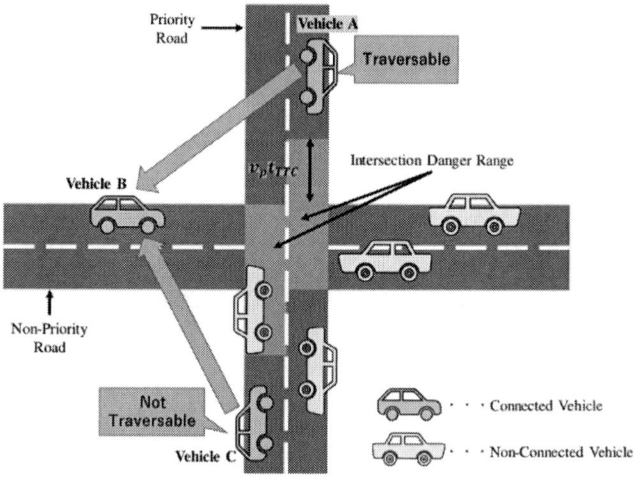

Figure 1. Usage example

We proposed a method that enables connected vehicles on priority (higher traffic volume) roads near an intersection to sense the presence of nearby vehicles (both connected and non-connected) and to then share with connected vehicles on non-priority (lower traffic volume) roads via V2V communication

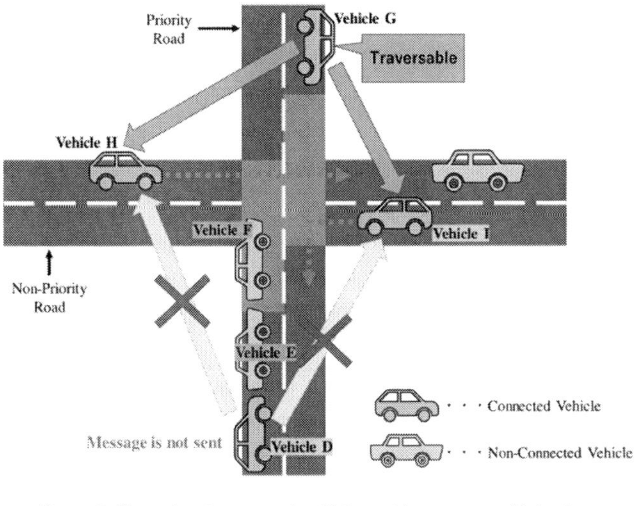

Figure 2. Example of connected vehicle unable to sense vehicles in intersection danger range

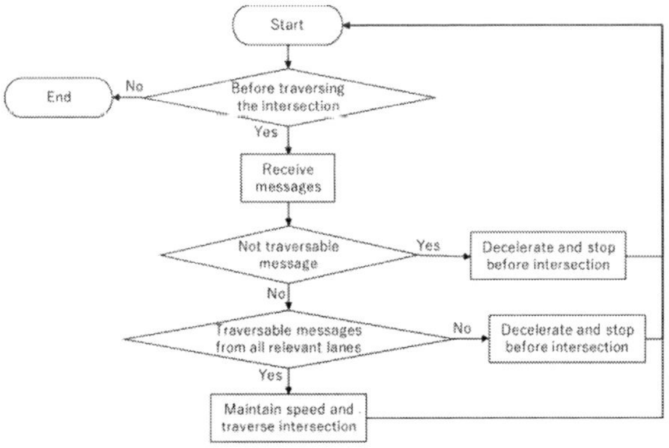

Figure 3. Judgement flow for connected vehicle on non-priority road about whether to enter intersection

information about whether they can enter and traverse the intersection. Using a traffic flow simulator, we considered safety and evaluated the relationship between traffic flow efficiency and the prevalence of connected vehicles when connected vehicles on non-priority roads approach and traverse an intersection.

The remainder of this paper is structured as follows. Section II details of related works. Section III details of the proposed method to share and use the information between connected vehicles in the mixed situation. Section IV details the evaluation of the proposed method. Section V details the results of the evaluation. Section VI details the consideration obtained from the evaluation results. Section VII details the conclusion.

II. RELATED WORK

In this section, we introduce two related works.

A. Collective Perception

Günther et al. [3] proposed a method for enabling connected vehicles to send environmental perception messages (EPM) to inform other connected vehicles of the position of surrounding vehicles detected by their radar sensors in addition to cooperative awareness messages (CAM) [4] to notify other connected vehicles of their existence in a mixed situation. It was shown that connected vehicles could perceive the positions of many vehicles within a radius of 300 meters (communicable range) around even if the percentage of connected vehicles among those vehicles (the penetration rate) was less than 100%.

B. Safety and Efficiency of Connected Vehicles Traversing an Intersection

Kimura et al. [5] proposed a method for enabling connected vehicles on a non-priority road to obtain the speed and current position of connected vehicles on a priority road using V2V communication and then use it to determine whether it is safe to enter an intersection without stopping to check for oncoming vehicles. It was shown that the travel time of vehicles on the non-priority road was lower than with two conventional methods: stopping before entering an intersection to visually check for approaching vehicles and using traffic lights.

Since this method is based on the premise that all vehicles are connected vehicles, a method is needed that takes into account the possible presence of non-connected vehicles as well.

III. PROPOSED METHOD

We first explain the communication procedure of the proposed method and then explain the operation procedure for connected vehicles. An example of using the proposed method is shown in Figure 1. For simplicity, each road has only two lanes. Here, v_p is the speed limit of the priority road, and t_{TTC} is the Time-To-Collision (TTC). These are declared in order to use for the explanation in subsection D.

A. Communication Procedure

The intersection and a portion of the lanes leading into the intersection are defined as the "intersection danger range." If one or more vehicles on the priority road are within this range, it is judged that it is dangerous for vehicles on the non-priority road to enter the intersection. Connected vehicles on the priority road sense this intersection danger range (either in front or behind them). If they do not sense any vehicles in this range, they broadcast a message saying that the intersection is traversable, meaning that connected vehicles on the non-priority road approaching the intersection can safely enter the intersection without stopping to check for oncoming vehicles. If they sense one or more vehicles in this range, they broadcast a message saying that the intersection is not traversable, meaning that connected vehicles on the non-priority road approaching the intersection cannot safely enter the intersection. As illustrated in Figure 1, the connected vehicle traveling from top to bottom (Vehicle A) does not sense any vehicles in the intersection danger range ahead and transmits a "traversable message." The connected vehicle traveling from bottom to top (Vehicle C) does sense a vehicle in the danger range and transmits a "not traversable message."

The connected vehicle on the non-priority road enters the intersection without stopping only if traversable messages are received from all relevant lanes on the priority road. The relevant lanes are those intersecting the trajectory of the vehicle entering the intersection from the non-priority road. If a not traversable message is received, like that vehicle B in Figure 1, or if a traversable message for all relevant lanes is not received, the driver decelerates and stops before the intersection.

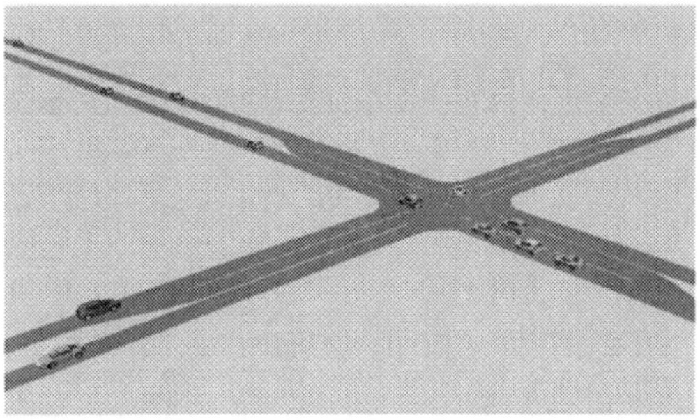

Figure 4. Execution screen of Vissim

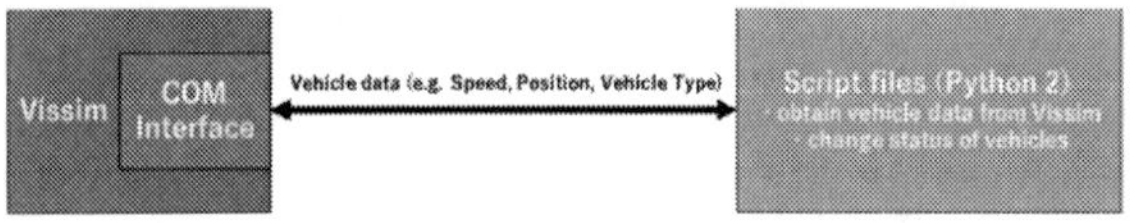

Figure 5. Configuration of Vissim

B. Operation Procedure for Connected Vehicles on Priority Road

If a connected vehicle is in a position where it can sense vehicles in the intersection danger range, either in front or behind it, it sends either a traversable or not traversable message depending on the sensed situation. Otherwise, no message is sent.

- If it does not detect any vehicles in the intersection danger range, it transmits a traversable message.
- If it detects one or more vehicles in the intersection danger range, it transmits a not traversable message.
- If the connected vehicle itself is in the intersection danger range, it transmits a not traversable message.
- If it cannot sense vehicles in the intersection danger range due to other vehicles in front or behind, no message is sent.

Figure 2 shows an example of the last situation. The connected vehicle running on the priority road from bottom to top (Vehicle D) is unable to sense vehicles in the intersection danger range (in this case vehicle F) due to the presence of a vehicle ahead of it (Vehicle E). Vehicle D thus does not transmit a message. This prevents vehicles in the intersection danger range from being overlooked.

C. Operation Procedure for Connected Vehicles on Non-Priority Road

A connected vehicle on a non-priority road approaching an intersection with a priority road constantly receives messages from connected vehicles on the priority road that are within communication range. The connected vehicle uses these messages to determine whether to enter the intersection without stopping to check for oncoming vehicles. The judgement flow is shown in Figure 3.

- If traversable messages are received from all relevant lanes on the priority road and a not traversable message is not received, the connected vehicle enters the intersection without stopping.
- Otherwise, the connected vehicle decelerates and stops before the intersection.

The situation in which a connected vehicle on a non-priority road does not receive traversable messages for all relevant lanes is illustrated in Figure 2. The connected vehicle running on the non-priority road from left to right (Vehicle H) intends to proceed straight through the intersection. A traversable message is received from vehicle G but not from vehicle D. Since the connected vehicle cannot confirm the safety of the intersection, it does not enter the intersection without stopping. On the other hand, the connected vehicle running on the non-priority road from right to left (Vehicle I) intends to turn left. Again, a traversable message is received from vehicle G but not from vehicle D. However, the connected vehicle can enter the intersection because the vehicles' trajectory passes only through vehicle G's lane.

Incidentally, non-connected vehicles on the non-priority road always stop before the intersection to check the safety of the intersection as in conventional intersections with stop signs.

D. Safety with Intersection Danger Range

After defining the intersection danger range, we consider the safety of the situation. The intersection danger range is the range in which a vehicle on a non-priority road may collide with a vehicle on the priority road upon entering their intersection. Its length L is calculated using the speed limit on the priority road, v_p, and the Time-To-Collision (TTC), t_{TTC}:

$$L = v_p t_{TTC} \tag{1}$$

We assume that connected vehicles can be human-operated vehicles in which information is notified to drivers through on-board equipment and drivers make decisions and perform operations, as well as autonomous vehicles. Thus, we defined safety as not only the prevention of collisions at intersections but also as the reassurance of drivers of human-operated vehicles about the behavior of autonomous vehicles when both types are on the same road. Drivers on a priority road may actually be surprised by autonomous vehicles entering an intersection from a non-priority road and brake suddenly. This may affect trailing vehicles and lead to traffic jams and collisions. The TTC, a parameter in determining whether a connected vehicle on a non-priority road enters an intersection without stopping, must include a time margin prevent surprising drivers of vehicles on the priority road.

Therefore, the minimum TTC for connected vehicles on priority roads to transmit traversable messages differ between autonomous and human-operated vehicles. While the TTC for autonomous vehicles was set to the maximum time required for a vehicle on a non-priority road to traverse an intersection, that for human-operated vehicles was set sufficiently higher to prevent drivers from being surprised. Varying the TTC enables both efficiency and safety to be achieved compared with a fixed TTC. We obtained the time margin from a study that analyzed the relationship between the TTC for a pedestrian and the driver's surprise when a pedestrian suddenly started crossing the road [6]. Although the target was a pedestrian, the situation is similar to that of vehicles entering from intersecting roads.

IV. EVALUATION

In this section, we explain the simulation environment and the way to evaluate efficiency of our proposed method.

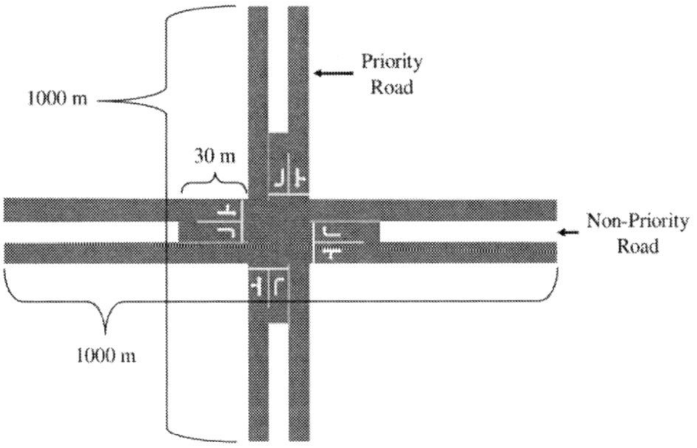

Figure 6. Layout of intersection used for evaluation simulation

TABLE I. SIMULATION PARAMETERS

Parameter	Setting
Speed limit	Priority Road: 50 km/h Non-Priority Road: 40 km/h
Number of vehicles per lane per hour	50 - 550
Penetration rate of connected vehicles	0 % - 100 %
Ratio of vehicles (Priority:Non-Priority)	3:1
Lane width	3.5 m
Measurement time	30 minutes
Number of measurements	10
Measurement section	530 m
Minimum TTC to transmit traversable message	Autonomous: 3.5 s Human-operated: 5.0 s (3.5 s + 1.5 s time margin)

A. Connected Vehicles

The connected vehicles in our simulation evaluation were assumed to satisfy the following conditions. The communication range and communication frequency were in accordance with the ITS communication requirements of the Japanese Ministry of Internal Affairs and Communications [7], the European Telecommunications Standards Institute standards [4], and the Society of Automotive Engineers standards [8]. The radar sensing range matched that of the in-vehicle mm-wave radar now in practical use [9].

- Each connected vehicle can communicate with other connected vehicles within a radius of 250 m.
- The communication frequency is 100 ms.
- The connected vehicles are equipped with a radar sensor that can detect a vehicle 200 m in front or behind.

B. Simulator

We used Vissim [10], a microscopic multi-modal traffic flow simulator developed by Planung Transport Verkehr AG

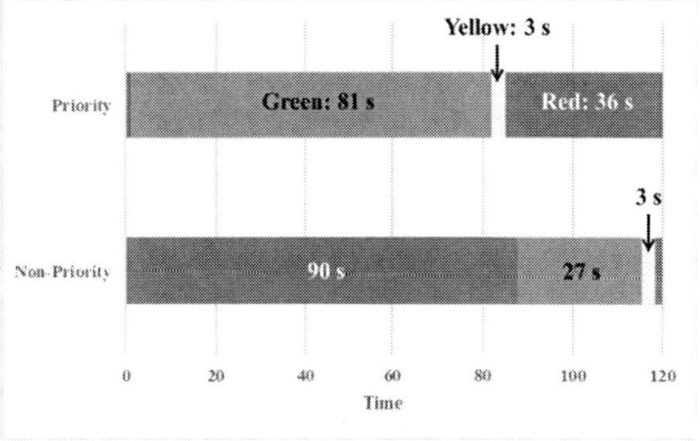

Figure 7. Setting of traffic lights

in Karlsruhe, Germany. As shown in Figure 4, Vissim can model various realistic road environments and visualize traffic phenomena with 3D graphics.

Vissim also supports the Component Object Model interface, and, as shown in Figure 5, can read script files by using this interface. The vehicle data in Vissim was obtained using script files programmed in Python 2, and the operation of the connected vehicles was described on the basis of that data.

C. Evaluation Setting

The simulated environment (Figure 6) was a single intersection between two roads, each with a length of 1000 m, that intersect at the midpoint of each. The roads have the same lane width and number of lanes; one was designated as the priority road.

We measured the travel time delay (TTD) and the maximum queue length (MQL) on the non-priority road and used them as evaluation indexes of efficiency. Travel time is the average time taken for a vehicle to traverse a specific section. We set the measurement section length to 530 m, which is the length from the starting point of the road to the point where the intersection ended. Since the actual travel time depends on the length of the measurement section, we also measured the ideal travel time, i.e., the time it takes to traverse the same length without stopping when entering an intersection. We defined the difference between the actual time and the ideal time as the TTD. The MQL is the maximum length of the traffic queue at the intersection.

These indexes were evaluated by changing the number of vehicles per lane per hour (traffic volume) and the penetration rate of connected vehicles. Table I lists the parameter settings. The speed limit and traffic volume were set in accordance with the typical conditions for roads in Japan [11] [12]. The simulation runtime was 30 minutes, and there were ten runs. The results for runs were averaged.

D. Comparison with Conventional Methods

For comparison purposes, we created models of two conventional methods: stopping before entering an intersection to visually check for approaching vehicles (stop model) and using traffic lights (traffic light model). They were evaluated under the same conditions.

The stop model is the conventional intersection with stop signs. All vehicles on the non-priority road stop before the stop sign for 0.5 s and then enter it after determining that it is safe to do so.

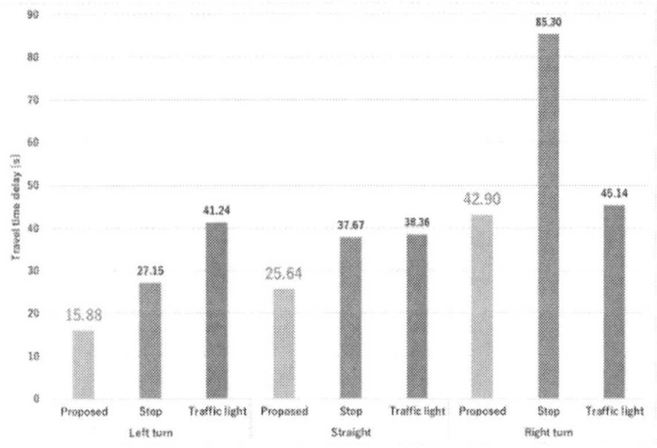

Figure 8. Comparison of travel time delay with model

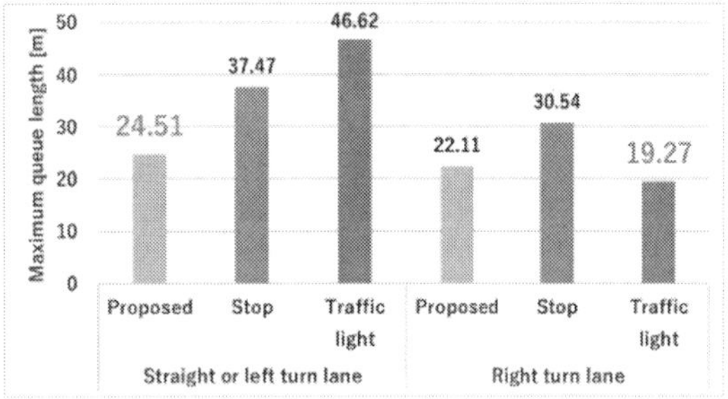

Figure 9. Comparison of maximum queue length with model

The traffic light model is the conventional intersection with traffic lights. All vehicles obey the traffic lights. Figure 7 shows the traffic light settings. The cycle time, i.e., the time required for a the traffic light to cycle from green to yellow to red, was set to 120 s, and the durations of the red and green lights were determined on basis of the traffic volume ratio between the priority and non-priority roads.

V. RESULTS

We show the results of the above simulation experiment.

A. Comparison of TTD and MQL Between Proposed Method and Conventional Models

Figures 8 and 9 respectively show the results for TTD and MQL when the traffic volume was 500 vehicles per hour and the penetration rate of connected vehicles was 70%.

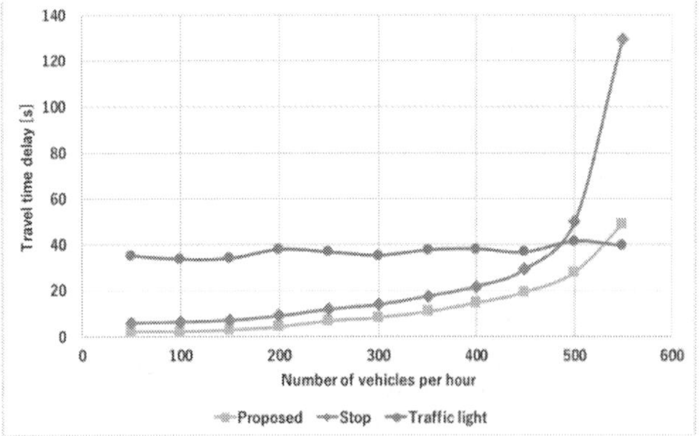

Figure 10. Changes in travel time delay with traffic volume

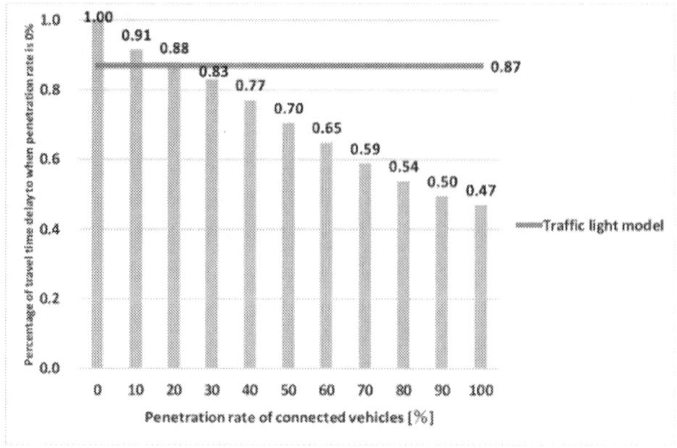

Figure 11. Decreasing rate in travel time delay against penetration rate of connected vehicles

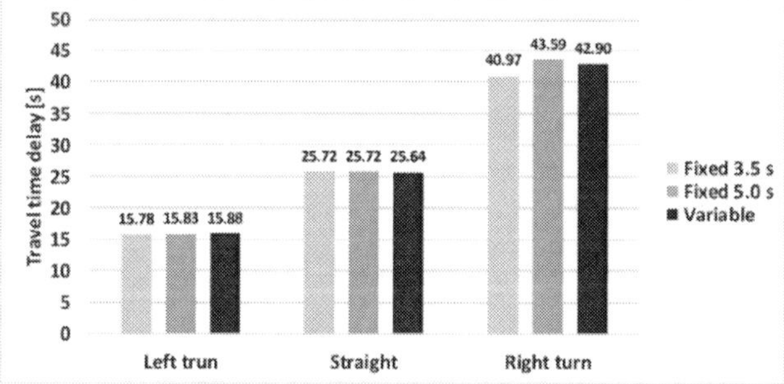

Figure 12. Changes in travel time delay with TTC

The TTD in Figure 8 represents the difference between the actual travel time and ideal travel time for each vehicle that made a left turn, right turn, or proceeded straight ahead. For all turning patterns, the travel time with the proposed method (yellow) was the shortest. In particular, the TTD for left turn and straight ahead were reduced compared with the stop model (gray) and traffic light model (green). For right turn, the travel time with the proposed method was half that with the stop model and not significantly different from that with the traffic light model.

For the straight and left turn lanes, the MQL with the proposed method was the shortest, whereas for the right turn lane, it was longer than with the traffic light model, as shown in Figure 9. To make a right turn with the proposed method, messages from connected vehicles in both lanes of the priority road and the opposite lane of the non-priority road are needed, which is assumed to have made the queue longer than that with the traffic light model. (Note that traffic in Japan runs on the left, so making a right turn requires crossing the opposite lane.)

B. Change in TTD with Traffic Volume

Figure 10 shows the changes in average TTD for all turning patterns when the penetration rate was 70% and the traffic volume on the priority road was increased from 50 to 550 vehicles per hour in steps of 50.

The TTD with the proposed method and stop model gradually increased as the traffic volume increased while it remained almost constant with the traffic light model. When the traffic volume was 500 vehicles per hour or less, the TTD with our method was the shortest. When the traffic volume was 550 vehicles per hour or more, it was the shortest with the traffic light model.

C. Decrease in TTD with Penetration Rate

Figure 11 shows the results of TTD when the traffic volume on the priority road was 500 vehicles per hour and the penetration rate of connected vehicles was increased from 0% to 100% in steps of 10%. It shows the TTD for each penetration rate, as a percentage of the TTD when the penetration rate was 0% (equal to the value of the stop model). The green line shows the result with the traffic light model under the same conditions.

The TTD decreased monotonically as the penetration rate increased. When the penetration rate was 50%, the TTD was 30% lower than when the rate was 0%, and when the rate was 90%, it was 50% lower. The TTD was less than that with the traffic light model when the penetration rate of 30% or more.

D. Comparison of TTD with Fixed Versus Variable TTC

Figure 12 shows the TTD when the traffic volume was 500 vehicles per hour, the penetration rate of connected vehicles was 70%, and the minimum TTC for connected vehicles on the priority road to transmit traversable message was one of three patterns. The first pattern was taken from the proposed method: set the TTC to 3.5 s if the connected vehicle is an autonomous driving vehicle and set it to 5.0 s if it is human-operated vehicle (variable TTC). The second pattern is to set it to 3.5 s for all vehicles (fixed 3.5 s TTC), and the third is to set it to 5.0s for all vehicles (fixed 5.0 s TTC).

The differences in the TTD among the three patterns were small for left turn and straight ahead. For right turn, the delay with variable TTC was more than that with fixed 3.5 s TTC, and less than that with fixed 5.0 s TTC. As described above, fixed 3.5 s TTC would not be safe in a situation with a mixture of autonomous and human-operated vehicles. In short, variable TTC is more efficient than fixed 5.0 s TTC and safer than fixed 3.5 s TTC.

VI. DISCUSSION

We summarize our discussion according to the results of Section V.

A. Advantages of Proposed Method

With the proposed method, TTD and MQL on the non-priority road decreased compared with the conventional method of stopping before the intersection and then entering it after determining that it is safe to do so. If traffic volume is about 500 vehicles per hour, our method is more efficient than using traffic lights. In Japan, the average of the traffic volume is 440 vehicles per hour [12]. Therefore, the proposed method is effective at intersections with an average traffic volume. Moreover, it is effective even during the early stages of connected vehicles introduction because its efficiency is better than that using traffic lights when the penetration rate is 30% or more.

Furthermore, the proposed method does not require mediation devices such as traffic lights and roadside devices because it used only V2V communication. Thus, the cost of device installation and maintenance is eliminated.

In situations where there are both autonomous and human-operated vehicles, we found that safety can be ensured by setting the TTC sufficiently high to prevent surprising drivers of vehicles on the priority road. It is also possible to change the TTC more dynamically in accordance with other characteristics such as driver's age and vehicle type. In this study, only connected vehicles on the priority road judged whether it was safe for vehicles on the non-priority road to enter the intersection. A more advanced method would be to have the connected vehicles on the priority road transmit the TTC information to the connected vehicles on the non-priority road. This would enable a connected vehicle on the non-priority roads to take into account the vehicle's characteristics when judging whether it is safe for the vehicle to enter the intersection.

B. Disadvantages of the Proposed Method

This method would not work at intersections with heavy traffic on the priority road, such as many intersections in urban areas, because there are normally few breaks in the traffic flow that would allow vehicles on the non-priority road to enter the intersection. Another method is needed, such as the traffic lights method or a method in which connected vehicles on the non-priority road could transmit an entry request to connected vehicles on the priority road.

VII. CONCLUSION

Looking ahead to the time when connected and non-connected vehicles will share the road, we developed a method that enables connected vehicles on a non-priority road to traverse an intersection with a priority road more quickly on the basis of information in messages from connected vehicles on the priority road. An evaluation simulation designed to examine the safety and efficiency of connected vehicles on a non-priority road traversing an intersection showed that efficiency can be improved, and that safety can be ensured. We need further research and develop another method to make intersections with heavy traffic efficiency in the mixed situation.

REFERENCES

[1] Ministry of Internal Affairs and Communications of Japan, "ITS initiatives in Japan", 2013, URL: https://www.mlit.go.jp/road/ITS/pdf/ITSinitiativesinJapan.pdf [accessed: 2019-12-23].

[2] Reuters, "Toyota abandons plan to install U.S connected vehicle tech by 2021", 2019, URL: https://www.reuters.com/article/us-autos-toyota-communication/toyota-abandons-plan-to-install-u-s-connected-vehicle-tech-by-2021idUSKCN1S2252 [accessed: 2019-12-23].

[3] H. Günther, R. Riebl, L. Wolf, and C. Facchi, "Collective Perception and Decentralized Congestion Controlin Vehicular Ad-hoc Networks", Vehicular Networking Conference(VNC), pp. 1-8, 2016.

[4] European Telecommunications Standards Institute (ETSI), "Intelligent Transport Systems (ITS); Vehicular Communications; Basic Set of Applications; Part 2: Specification of Cooperative Awareness Basic Service", 2019.

[5] K. Kimura, S. Azuma, and K. Sato, "Evaluation of Safety and Efficiency Simulation of Cooperative Automated Driving", The Seventh International Conference on Advances in Vehicular Systems, Technologies and Applications (VEHICULAR 2018), pp. 66-71, Venice, Italy, 2018.

[6] A. Nakamura, S. Tominaga, and M. Okano, "Car-to-pedestrian Hiyari-Hatto Incident Analysis by Using Drive-recorder", The 2010 Technical Papers of Academic Lecture, Faculty of Science and Engineering, Nihon University, pp. 343-344, 2010.

[7] Ministry of Internal Affairs and Communications of Japan, "Study Group Report on Advanced ITS Wireless Systems", 2009.

[8] SAE International, "On-Board System Requirements for V2V Safety Communications", 2016.

[9] Hitachi Automotive Systems, Ltd., "Hitachi Automotive Systems Develops World's Smallest Class 77GHz Millimeter-wave Radar as a Long-range, Forward-direction Sensor for Autonomous Driving", 2017, URL: https://www.hitachi.com/New/cnews/month/2017/10/171003.html [accessed: 2019-08-10].

[10] PTV Group, "PTV Vissim", URL: http://vision-traffic.ptvgroup.com/en-us/products/ptv-vissim/ [accessed: 2019-05-25].

[11] Ministry of Land, Infrastructure, Transport and Tourism of Japan, "Government Order on Road Design Standards", 2008, URL: https://www.mlit.go.jp/road/sign/kouzourei_kaisetsu.html [accessed: 2019-07-07].

[12] Ministry of Land, Infrastructure, Transport and Tourism of Japan, "National Road and Street Traffic Situation Survey", 2015, URL: https://www.mlit.go.jp/road/census/h27/index.html [accessed: 2019-07-07].

Towards an Integrated In-Vehicle Isolation and Resilience Framework for Connected Autonomous Vehicles

Khaled Mahbub, Mohammad Patwary, Antonio Nehme
Birmingham City University
Birmingham, United Kingdom
Email: {firstname.lastname}@bcu.ac.uk

Marc Lacoste, Sylvain Allio, Yvan Rafflé
Orange Labs
France
Email: {firstname.lastname}@orange.com

Abstract—**Connected Autonomous Vehicles (CAV) have attracted significant attention, specifically due to successful deployment of ultra-reliable low-latency communications with Fifth Generation (5G) wireless networks. Due to the safety-critical nature of CAV, reliability is one of the well-investigated areas of research. Security of in-vehicle communications is mandatory to achieve this goal. Unfortunately, existing research so far focused on in-vehicle isolation or resilience independently. This short paper presents the elements of an integrated in-vehicle isolation and resilience framework to attain a higher degree of reliability for CAV systems. The proposed framework architecture leverages benefits of Trusted Execution Environments to mitigate several classes of threats. The framework implementation is also mapped to the AUTOSAR open automotive standard.**

Keywords - Isolation; Resilience; ECU; Monitoring; Trusted Execution Environment; AUTOSAR; Certification.

I. INTRODUCTION

Despite considerable progress in the last decade, the development of fully self-driving vehicles is still largely under research and experimentation. In such safety-critical systems, the resilience of in-vehicle and inter-vehicle communication is a key element to ensure the security of the vehicle. While in-vehicle relates to on-board communication between Electronic Control Units (ECUs) acting based on inputs from different sensors, inter-vehicular communications enable data exchange with the external environment including other vehicles, broadband clouds and roadside-infrastructures [1].

In this system of systems model, the diversity, autonomy and connectivity of vehicles mean vulnerabilities at the level of the vehicle affect the larger environment [2]. While both types of communication enable safety-critical decision-making, in-vehicle communication requires special attention. The disparity of coding practices among the diversity of specialised vendors in different functionalities (e.g., infotainment, braking and steering assistance), and the trust model induced by the high degree of connectivity and unrestricted interactions between vehicle components to enable comfort features (e.g., adjusting the sound volume according to the velocity) widen the attack surface [1].

Internal security barriers to detect and react to an intrusion are therefore needed to limit the impact of a compromise and to mitigate its propagation to different subsystems within a vehicle [1]. Moreover, the adoption of new technologies in the automotive domain is opening new safety and security challenges. For example, the advent of new generations of ECUs that are virtualized as lightweight execution environments (e.g., virtual machines, containers) on different types of virtualization platforms, (e.g., OKL4 Microvisor, Proteus Hypervisor, ETAS STA-HVR [3]) may face system-level isolation challenges such as side-channels.

This short paper introduces our approach to detect and limit the impact of intrusions for in-vehicle networks that can compromise the safety of autonomous vehicles. This will be a step towards enhancing the robustness of in-vehicle communications through the isolation of ECUs, the detection of and recovery from intrusions. Focusing on spoofing, replay, and side-channel attacks, we present principles of a framework for in-vehicle isolation and resilience and discuss technical considerations for its implementation according to the AUTomotive Open System ARchitecture (AUTOSAR) open standard.

This paper is structured as follows: Section II presents related work. Section III introduces our framework and Section IV discusses considerations to adhere to the AUTOSAR standard. We conclude our paper in Section V.

II. RELATED WORK

A significant body of work focuses on improving resilience of connected autonomous vehicles. Solutions against threats can be categorised as i) Proactive Defence, ii) Active Defence and iii) Passive Defence [4]-[8]. We give next a brief overview of each family of techniques.

A. Proactive Defence

Proactive defence is underpinned by the "security by design" principle practiced in the software industry [6],[7]. Integration of common security practices, public key encryption and hash-based message authentication fall under this category [4],[9].

B. Active Defence

Active defence mitigates threats as they occur. For instance, continuous monitoring can be applied to detect intrusions and preserve the security hygiene of the vehicle and take adequate remediation actions [10]; in this sense, real-time monitoring enables the identification and isolation of faulty applications in safety-critical systems [11].

Detection approaches for the in-vehicle network can be categorised as i) Signature-Based Detection, ii) Anomaly-Based Detection and iii) Hybrid Approach [12]-[15]:

- Signature-Based Detection: These approaches use information about attacks (signatures) as a pattern characterizing known threats, comparing signatures against observed events to identify possible attacks.
- Anomaly-Based Detection: These approaches are based on continuous monitoring of system activities, checking against a reference model (e.g., profile of the system). An alarm is raised if deviation from the reference model is observed. Various mechanisms can be applied to derive the reference model, such as machine learning [16],[17], frequency-based [18]-[20], and statistical-based methods [21],[22].
- Hybrid Approach: This family of approach comprises several intrusion detection techniques (e.g., signature- and anomaly-based detection).

In addition to in-vehicle intrusion detection, several approaches explore detection of side channel attacks for the automotive domain - at the physical layer [23], using cache-based [24] or interface-based approaches [25],[26].

C. Passive Defence

Passive defence mainly focuses on detecting, responding to, and recovering from an attack once it has occurred. This type of defence is notably suitable to prevent malwares and code injection and modification threats. Therefore, passive defences are not suitable for safety-critical systems, like autonomous vehicles, as these approaches do not facilitate detection and mitigation of adversaries in real-time [8],[10].

It should be noted that proactive defence and passive defence are not suitable to handle adaptive security requirements, very common in the cyber and the automotive domains. Proactive defence recommends designing control features to meet the security objectives at system design time and embedding such features in the system. However, this approach is unable to cover new types of threats once the system has been developed. On the other hand, passive defences alone are not suitable for safety-critical systems, such as autonomous vehicles, as these approaches detect the attack once it has occurred. Also the active defence techniques approaches found in the literature apply continuous monitoring to detect anomalies, but did not consider their application to secure execution environments for ECUs. As described in the next sections, our approach aims to address these limitations.

III. In-Vehicle Isolation and Resilience

We adopt the active defence approach to improve in-vehicle resilience: security properties related to the communication among ECUs will be continuously monitored in order to detect security threats, and actions will be taken to mitigate the impact of and gracefully recover from the detected threat. Recovery in our context consists of rolling back (or forward) to a stable state to overcome intrusions [27].

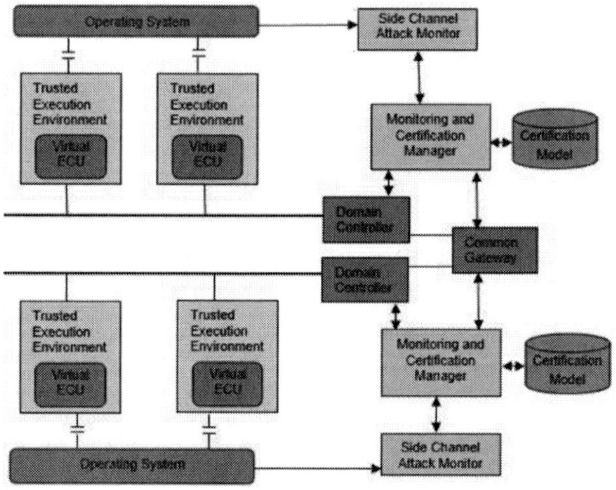

Figure 1. Reference Architecture of Isolation and Resilience Framework

Figure 1 shows the proposed reference architecture for threat detection and mitigation in the in-vehicle network. ECUs are grouped into different domains according to similarity of functionalities. All ECUs in a domain are connected to the same communication bus and activities of each ECU in a domain are controlled by one domain controller. Domain controllers are connected through a common gateway in order to enable communication among the ECUs belong to different domains [4],[9]. The major components of the architecture are:

A. Trusted Execution Environment (TEE)

Trusted Execution Environments enable to specify isolated execution environment in the main processor [28],[29]. The TEE provides security features such as isolated execution, integrity of applications executing in the TEE, and confidentiality of application assets. Several hardware vendors provide embedded technologies that can be used to support TEE implementations, including AMD PSP [30], ARM TrustZone [31], EVITA Hardware Security Modules (HSM) [32] and Intel SGX Software Guard Extensions [33]. We aim to explore if TEEs could be applied as secure execution environment for ECUs, thereby ensuring secure/isolated communication from ECU to ECU.

B. Side Channel Attack Monitor

While TEEs aim to provide secured execution environments, they are vulnerable to side-channel attacks [34],[35]. This component focuses on runtime detection of the variants of side-channel attacks (e.g., SGX interface-based attacks) that are relevant in a vehicular context.

C. Monitoring and Certification Manager

The responsibility of this component is to perform real-time monitoring of security properties related to components (e.g., ECUs) in the in-vehicle network to detect security threats. This component applies the hybrid approach (including frequency-based, statistical-based, and deep packet inspection approaches) to detect spoofing and replay attacks. Based on the validity of the security properties, this component also maintains the certificates (detailed in Section IV.B) that certify the valid state of the ECUs.

IV. Implementation Considerations

We adopt the AUTOSAR open standard for automotive software architecture and framework to implement the architecture presented in Section III. The AUTOSAR consortium was formed by major automotive OEMs like BMW, Ford, Daimler and Chrysler to standardize the automotive software architecture and framework, thereby facilitating scalability, reusability and interoperability across the products lines from different OEMs [36]. The use of AUTOSAR in the implementation would therefore inherently enable the prototype to have the same benefits.

Next, we briefly introduce AUTOSAR, and then discuss the mapping between our framework and AUTOSAR. In AUTOSAR, the ECU software is abstracted in a layered architecture, built on top of the underlying micro-controller hardware [37]. As shown in Figure 2, this architecture is composed of three layers, namely Basic Software (BSW), Runtime Environment (RTE), and Application Layer.

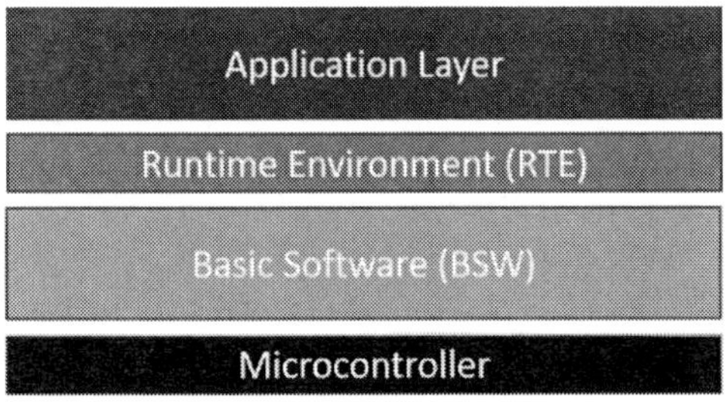

Figure 2. Overview of AUTOSAR components [37]

Basic Software Layer (BSW) is the bottom layer of the architecture and provides core system functionalities. This layer has 3 sub-layers. First, the Micro-controller Abstraction Layer (MCAL) contains internal driver modules that access the underlying micro-controller and internal peripherals directly. Second, the ECU Abstraction Layer (ECUAL) interfaces the drivers of the MCAL and offers an API for accessing the peripherals and external devices, thus making higher software layers independent of the hardware layout. And third, the Services Layer (SL) provides top-level services (e.g., operating system functionality, communication services, management services, memory services, etc.) to application software components.

The Run-Time Environment (RTE) layer provides communication services to the application software, acting as a bridge between the application and the BSW layer.

The Application Layer mainly consists of software components (SWC) interconnected to other SWCs and BSW modules. This layer is component-based, which enhances SWC scalability and re-usability.

Figure 3 shows the mapping of the major components of our framework to the AUTOSAR architecture. We propose to add an ECU that takes the role of monitoring existing ECUs in the system, and to isolate ECUs.

The left side (yellow box) of the Figure 3 shows the deployment of virtual ECUs within the TEE, following the AUTOSAR architecture. The right side of the Figure 3 shows the Domain Controller, Monitoring & Certification Manager and Side Channel Attack Monitor components of the framework developed as SWCs in the AUTOSAR application layer, i.e., these software components will reside within a trusted virtual ECU and will collect the data transmitted among the virtual ECUs of the in-vehicle network. Such data will be used for monitoring the security properties related to different ECUs.

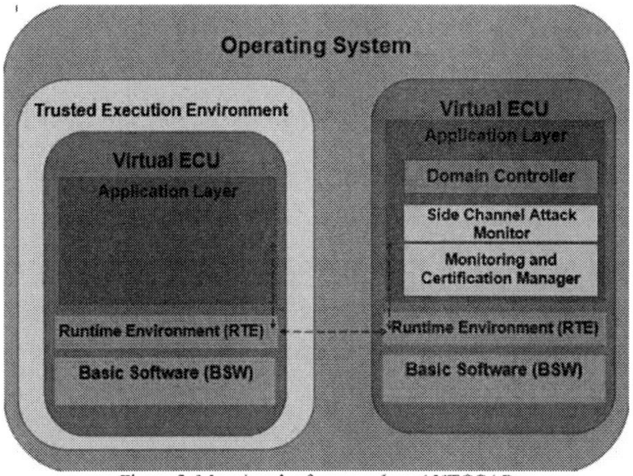

Figure 3. Mapping the framework to AUTOSAR

A. Monitoring

As shown in Figure 3, the Monitoring and Certification Manager and the Side Channel Attack Monitor will collect the data transmitted among the virtual ECUs of the in-vehicle network. A sub-component, namely *DataCollector*, will be deployed for transforming the data from a legacy format (e.g., CAN bus, being the most widely used protocol in the automotive industry [38]) to a format that used for monitoring.

This design may help to support multiple communication standards in the framework (e.g., Automotive Ethernet) by implementing a dedicated *DataCollector* (e.g., converting in-vehicle data from Automotive Ethernet to network monitoring format). Using multiple monitoring ECUs (e.g., for each sets of Domain Controllers) may help addressing safety constraints to avoid single points of failure.

B. Certificate Model

The Monitoring & Certification Manager and Side Channel Attack Monitor perform real-time monitoring of security properties related to ECUs to detect security threats and produce a certificate for the in-vehicle network. Monitoring is driven by security properties expressed as condition-action rules verified by a rule engine (e.g., CLIPS [39]) against runtime facts (i.e., runtime data). Monitoring results are accumulated to produce a certificate that certifies the state of the monitored components (e.g., ECUs).

The certificate structure includes the following elements:

1) CertificateID: represents the unique identifier of a generated certificate during the monitoring process.

2) MonitoringResultAggregator: aggregates monitoring results produced by monitoring different components (e.g., ECUs, CAN bus etc.) of the in-vehicle network. This element contains the following sub-elements:

- *AggregationTime:* denotes the time of aggregation.
- *Duration:* specifies the timespan between monitoring results considered for aggregation.
- *ToMLis:* specifies a list of TargetOfMonitoring considered for the aggregation operation.
- *AggregationRule:* defines how monitoring results should be aggregated, e.g., for results with numerical values by applying statistical methods.
- *AggregationResult:* stores the aggregation result.

3) TargetOfMonitoring (ToM): a component (e.g., ECU, CAN bus, etc.) monitored to identify security threats associated with it.

The *ToM* has the following sub-elements

- *ToMType:* the type of component to be monitored.
- *ToMID:* a unique identifier of the component in the in-vehicle network.
- *MonitoringRule:* the security property related to this component to be monitored.
- *MonitoringEvidenceAggregator:* contains the aggregation of results by monitoring the *MonitoringRule* related to this component.

The *MonitoringEvidence Aggregator* contains the following sub-elements:

- *AggregationTime;* denotes the time of aggregation.
- *Duration:* specifies the time span between in-vehicle network data considered for monitoring.
- *AggregationRule:* defines how monitoring results should be aggregated, e.g., for results with numerical values by applying statistical methods.
- *AggregationResult:* stores the aggregation result.

V. CONCLUSION AND OUTLOOK

This position paper provided an overview of defence strategies to mitigate common threats to in-vehicle networks. We proposed an architecture and framework to enhance in-vehicle isolation and resilience focusing on spoofing, replay and side-channel attacks. The framework follows an active defence strategy to detect and react to intrusions on the in-vehicle network and to recover from attacks. This recovery may be rolling back to a stable state to overcome an intrusion (e.g., in [27]), or to estimate the stable state by applying different techniques (e.g., in [5]). This framework may also be used to detect anomaly or misbehavior, which are not necessarily resulting from cyberattacks but simply from system faults and to limit their propagation in such a system of systems (e.g., in [40]).

Next steps are to implement the framework, and to evaluate its isolation and resilience benefits in a simple setting, first using simulations, before a possible testbed implementation.

REFERENCES

[1] M. Faezipour, M. Nourani, A. Saeed, and S Addepalli, "Progress and challenges in intelligent vehicle area networks," Communications of the ACM, vol. 55, no. 2, pp. 90-100, Feb. 2012.

[2] J. Boardman and B. Sauser, "System of Systems-the meaning of," 2006 IEEE/SMC International Conference on System of Systems Engineering, pp. 6-pp, Apr. 2006, IEEE.

[3] A.K. Rajan, A. Feucht, L. Gamer, and I. Smaili, "Hypervisor for Consolidating Real-Time Automotive Control Units: Its Procedure, Implications and Hidden Pitfalls," Journal of Systems Architecture, vol. 82, pp. 37-48, Jan. 2018.

[4] K. Daimi, M. Saed, S. Bone, and John Robb, "Securing Vehicle's Electronic Control Units," Twelfth International Conference on Networking and Services, 2016.

[5] V. Marquis et al., "Toward attack-resilient state estimation and control of autonomous cyber-physical systems," Systems and Information Engineering Design Symposium (SIEDS), pp. 70-75, 2018.

[6] D. A. Brown et al. "Automotive Security Best Practices: Recommendations for security and privacy in the era of the next-generation car," McAfee White Paper, Aug. 2016.

[7] A. Chattopadhyay and K. Lam, "Autonomous Vehicle: Security by Design," Oct 2018, arXiv:1810.00545v1 [Retrieved: 13-05-2020].

[8] M. Saed, K. Daimi, and S. Bayan, "A Survey of Autonomous Vehicle Technology and Security," VEHICULAR 2019.

[9] M. S. Ul Alam, "Securing Vehicle Electronic Control Unit (ECU) Communications and Stored Data," Master of Science Thesis, School of Computing, Queen's University Kingston, Ontario, Canada, Sep. 2018.

[10] V. L. Thing and J. Wu, "Autonomous Vehicle Security: A Taxonomy of Attacks and Defences," 2016 IEEE International Conference on Internet of Things (iThings) and IEEE Green Computing and Communications (GreenCom) and IEEE Cyber, Physical and Social Computing (CPSCom) and IEEE Smart Data (SmartData), pp. 164-170, Dec. 2016.

[11] B. Motruk, J. Diemer, R. Buchty, R. Ernst, and M. Berekovic, "IDAMC: A many-core platform with run-time monitoring for mixed-criticality," 2012 IEEE 14th International Symposium on High-Assurance Systems Engineering, pp. 24-31, Oct. 2012, IEEE.

[12] G. Dupont, J. Hartog, S. Etalle, and A. Lekidis. (2019). "Network intrusion detection systems for in-vehicle network." Technical report, https://arxiv.org/abs/1905.11587 [Retrieved: 13-05-2020]

[13] S. F. Lokman, A. T. Othman, and M. H. Abu-Bakar, "Intrusion detection system for automotive Controller Area Network (CAN) bus system: a review," EURASIP Journal on Wireless Communications and Networking. 2019, doi: 10.1186/s13638-019-1484-3.

[14] P Kaur, M. Kumar, and A. Bhandari "A review of detection approaches for distributed denial of service attacks" Systems Science & Control Engineering. pp. 301-320, Dec. 2016.

[15] A. Tomlinson, J. Bryans, and S. Shaikh, "Towards Viable Intrusion Detection Methods For The Automotive Controller Area Network," 2nd ACM Computer Science in Cars Symposium, pp. 1-9, Sep. 2018.

[16] E. Seo, H. M. Song, and H. K. Kim, "GIDS: GAN based Intrusion Detection System for In-Vehicle Network," 2018 16th Annual Conference on Privacy, Security and Trust (PST), pp. 1-6, 2018, doi: 10.1109/PST.2018.8514157.

[17] M. J. Kang and J. W. Kang, "Intrusion detection system using deep neural network for in-vehicle network security," PLoS One, vol. 11, no. 6, 2016.

[18] A. Taylor, N. Japkowicz, and S. Leblanc, "Frequency-based anomaly detection for the automotive CAN bus," 2015 World Congress on Industrial Control Systems Security (WCICSS), pp. 45-49, 2015, doi:10.1109/WCICSS.2015.7420322.

[19] C. Young, H. Olufowobi, G. Bloom, and J. Zambreno, "Automotive Intrusion Detection Based on Constant CAN Message Frequencies Across Vehicle Driving Modes," pp. 9-14, Mar. 2019, doi:10.1145/3309171.3309179.

[20] H. S. Sánchez, D. Rotondo, T. Escobet, V. Puig, J. Saludes, and J. Quevedo, "Detection of replay attacks in cyber-physical systems using a frequency-based signature," Journal of the Franklin Institute, vol. 356, no. 5, 2019.

[21] A. A. Sivasamy, and B. Sundan, "A dynamic intrusion detection system based on multivariate Hotelling's T2 statistics approach for network environments," The Scientific World Journal, 2015, doi:10.1155/2015/850153.

[22] A. Qayyum, M. H. Islam, and M. Jamil, "Taxonomy of statistical based anomaly detection techniques for intrusion detection," IEEE Symposium on Emerging Technologies, pp. 270-276, doi:10.1109/ICET.2005.1558893.

[23] S. Jain, Q. Wang, M. T. Arafin, and J. Guajardo, "Probing Attacks on Physical Layer Key Agreement for Automotive Controller Area Networks," Asian Hardware Oriented Security and Trust Symposium, pp. 7-12, 2018.

[24] Y. Kulah, B. Dincer, C. Yilmaz, and E. Savas, "SpyDetector: An approach for detecting side-channel attacks at runtime," International Journal of Information Security, vol. 18, pp. 393–422, doi.org/10.1007/s10207-018-0411-7.

[25] J. Wang, Y. Cheng, Q. Li, and Y. Jiang, "Interface-Based Side Channel Attack Against Intel SGX," Oct. 2018, https://arxiv.org/abs/1811.05378 [Retrieved: 13-05-2020]

[26] N. Weichbrodt, P. Aublin, and R. Kapitza, "sgx-perf: A Performance Analysis Tool for Intel SGX Enclaves," 19th International Middleware Conference, pp. 201-213, doi:10.1145/3274808.3274824.

[27] A. Binun, A. Bloch, S. Dolev, M. R. Kahil, B. Menuhin, R. Yagel, T. Coupaye, M. Lacoste, A. Wailly. "Self-stabilizing virtual machine hypervisor architecture for resilient cloud," 2014 IEEE World Congress on Services pp. 200-207, June 2014. IEEE.

[28] M. Sabt, M. Achemlal and A. Bouabdallah, "Trusted Execution Environment: What It is, and What It is Not," 2015 IEEE Trustcom/BigDataSE/ISPA, Helsinki, pp. 57-64, 2015 doi:10.1109/Trustcom.2015.357.

[29] "Trusted Execution Environment (TEE) 101: A Primer", Secure Technology Alliance, White Paper, Version 1.0, April 2018.

[30] AMD, AMD Secure Processor technology (AMD-SP), https://www.amd.com/en/technologies/security [Retrieved: 13-05-2020]

[31] "GlobalPlatform based Trusted Execution Environment and TrustZone-Ready: The foundations for trusted services", ARM, White Paper, October 2013.

[32] M. Wolf and T. Gendrullis, "Design, implementation, and evaluation of a vehicular hardware security module," International Conference on Information Security and Cryptology, pp. 302-318, Nov. 2011.

[33] J.-E. Ekberg, K. Kostiainen, and N. Asokan, "Trusted Execution Environments on Mobile Devices," ACM CCS 2013 Tutorial, https://www.cs.helsinki.fi/group/secures/CCS-tutorial/tutorial-slides.pdf [Retrieved : 13-05-2020].

[34] F. Brasser, U. Müller, A. Dmitrienko, K. Kostiainen, S. Capkun, and A. Sadeghi, "Software Grand Exposure: SGX Cache Attacks Are Practical," WOOT'17 Proceedings of the 11th USENIX Conference on Offensive Technologies, 2017.

[35] M. Schwarz, S. Weiser, D. Gruss, C. Maurice, and S. Mangard, "Malware Guard Extension: Using SGX to Conceal Cache Attacks," In: Polychronakis M., Meier M. (eds) Detection of Intrusions and Malware, and Vulnerability Assessment. DIMVA 2017. Lecture Notes in Computer Science, vol 10327. Springer.

[36] AUTOSAR History, https://www.autosar.org/about/history/, [Retrieved : 10-02-2020].

[37] AUTOSAR, "AUTOSAR: Layered Software Architecture.", https://www.autosar.org/fileadmin/user_upload/standards/classic/4-3/AUTOSAR_EXP_LayeredSoftwareArchitecture.pdf [Retrieved: 13-05-2020]

[38] C. Schlegel, "The role of CAN in the age of Ethernet and IoT," International CAN Conference (iCC), 2017.

[39] "CLIPS, A Tool for Building Expert Systems," http://www.clipsrules.net/index.html [Retrieved: 13-03-2020].

[40] A. Wasicek, M. D. Pesé, A. Weimerskirch, Y. Burakova, K. Singh "Context-aware Intrusion Detection in Automotive Control Systems," 5th ESCAR Conference, 2017.

Loss Performance of Intra-Vehicle Channels for Narrowband Signal Transmission

Irfan Yusoff, Xiao-Hong Peng and Jack Hydns

School of Engineering & Applied Science,
Aston University,
Birmingham, UK.
Email: mohdymni@aston.ac.uk, x-h.peng@aston.ac.uk

***Abstract*—In this paper, we examine the intra-vehicle radio propagation performance of narrowband signals at 2.4 GHz and 5.9 GHz. The measurements are taken from a vehicle testbed and analyzed based on the path loss model that comprises the mean, large-scale fading and small-scale fading loss components. We show that multipath fading, especially the small-scale fading, has the dominant impact on the loss performance, while the mean loss varies relative to the free-space loss in this environment. Different accumulated distribution functions are applied to assess their suitability for characterizing the large-scale and small-scale fading effects, and compared through the non-parametric tests.**

***Index Terms*—Intra-vehicle channel characteristics, path loss model, large-scale fading, small-scale fading, testbed.**

I. Introduction

Wireless communications have been introduced to vehicles, e.g.: Bluetooth and Wi-Fi are commonly used inside vehicles for infotainment, and the Global Navigation Satellite System (GNSS) is adopted for navigation. In addition, IEEE 802.11p and LTE-V2X systems [1] have been deployed to enable communication between vehicles (vehicle-to-vehicle or V2V) and between vehicles and roadside units (vehicle-to-infrastructure or V2I) for improving driving safety and data services to road users.

Applying wireless communication technologies inside a vehicle can improve its fuel and space efficiency through reducing wire harness which leads to the weight reduction and space saving for the vehicle. To achieve this end, the properties of intra-vehicle wireless channels need to be thoroughly investigated in order to develop appropriate technologies and protocols which can ensure the transmission performance required.

The aim of this work is to gain a better understanding of the characteristics of intra-vehicle channels for narrowband wireless signals in term of their distance-dependent attenuation factor and fading property at two different frequencies: 2.4 GHz and 5.9 GHz, which are chosen as they cover the 2.4 GHz unlicensed band and the 5.9 GHz dedicated short-range communications (or IEEE802.11p) band.

In addition, the investigation of the loss performance of the intra-vehicle channel is the focus of the empirical work we carried out. This will demonstrate that the overall path loss performance is contributed jointly by a mean loss component which follows the log-distance loss model, and multipath fading, namely large-scale and small-scale fading, related loss components. We will also show the level of influence from each of these components over the overall loss performance at different frequencies.

We have chosen to study the narrowband performance as it has been largely overlooked as compared to the published work on the Ultra-Wideband (UWB) transmission [2]. It is important to understand the narrowband behavior of the intra-vehicle channel as the wireless systems used or recommended for vehicles such as Bluetooth (no more than 2 MHz for each channel), Zigbee (2 MHz for each channel) and Wi-Fi (20 or 22 MHz for each channel) are all in this range. Furthermore, our results on the narrowband characteristics will provide a better understanding of the flat or deep fading behavior in the intra-vehicle environment.

Although there have been some investigations on the propagation characteristic of narrowband signals in intra-vehicle wireless channels, most of them are more focused at the general behavior of this type of channels, such as channel coherent bandwidth, frequency diversity and power delay profile. Some exemplar investigates include the work carried out by Cheng et al. which produces a simple analytical multi-ray model with field measurements [3], and the work by Liu et al. which analyses the potential benefit of frequency diversity for intra-vehicle wireless applications [4].

In addition, a study by Moghimim et. al. [5] examines channel coherence time and channel loss statistics at 915 MHz and 2.4 GHz but without considering the distance related loss performance. Kamoda et al. have presented both UWB and narrowband channel models for the engine compartment only which can be very different from the passenger compartment in propagation characteristics as they commented [6].

Our work provides more insight into the causes of intra-vehicle channel behavior by revealing the location/distance based loss performance and the fading phenomenon in both small scale and large-scale aspects and, at the same time, showing how they collectively contribute to the loss performance in such a complicated environment.

The structure of the paper is organized as follows. Section II describes the testbed used for this research and related settings based on which the measurements were collected. The results, including path loss and channel fading characteristics,

TABLE I
USRP SETTINGS AT 2.4 GHZ AND 5.9 GHZ

Frequency Setting	2.45 GHz	5.9 GHz
Antenna	ANT-2.4-LCW-SMA	TG.35.8113
Tx Power	-0.8 dBm	-5.18 dBm
Tx Gain Setting	73	73
Rx Gain Setting	35	60
Sample Rate	10Msps	10Msps
Update Rate	1Msps	1Msps

are presented in Section III together with proper discussions. Finally, the paper is concluded in Section IV.

II. TESTBED AND EXPERIMENT

A. Testbed Setup

The testbed was set up within the passenger and boot compartments in a Land Rover Discovery vehicle and two Universal Software Radio Peripheral (USRP) B210 devices from National Instruments were used as a transmitter and a receiver in the tests. Each USRP was connected to an omnidirectional antenna from Linx Technologies (2.4 GHz) and Taoglas Limited (5.9 GHz), and also attached to a laptop for signal generation and processing. The antennas were kept vertically polarised throughout the tests carried out with the transmitter and receiver aligned towards the back and front of the vehicle, respectively.

The transmitter was configured to transmit a baseband signal modulated by a carrier wave at the chosen frequency of 2.4 GHz or 5.9 GHz. The output power of the transmitter (USRP) was calibrated in reference to the component datasheet of the USRP and through a laboratory measurement with a Tektronix spectrum analyzer. The output power is set to be -1 dBm at 2.4 GHz and -5 dBm at 5.9 GHz. The receiver was configured to receive the complex signal waveforms for a length of 10 seconds, which resulted in a total of 100 million received samples in the time domain for each testing location. The measurements were collected through three runs for each test. The related settings are summarized and specified in Table I.

B. Measurement Procedure and Scenario

A frequency spectrum scan was performed before taking measurements to ensure the chosen channel to be free from interference. The vehicle was in a stationary position with the engine and electric power turned off during the tests. The area around the vehicle was also kept out of any large objects and human movement throughout the tests.

The transmitter was placed at a fixed location on the dashboard of the vehicle, while the location of the receiver (USRP) was changeable depending on the requirement in different tests. Two types of tests were conducted:

a) Test-1: Measuring the received signal power at 44 different locations across the passenger and boot compartments, as shown in Fig. 1, at 2.4 GHz and 5.9 GHz.

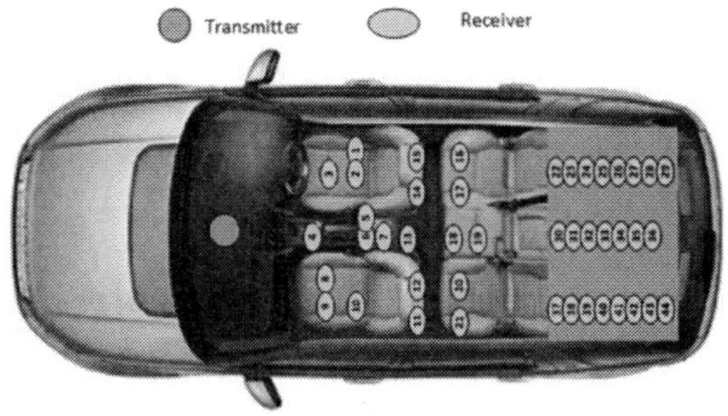

Fig. 1. Locations of 44 measurement points in Test-1.

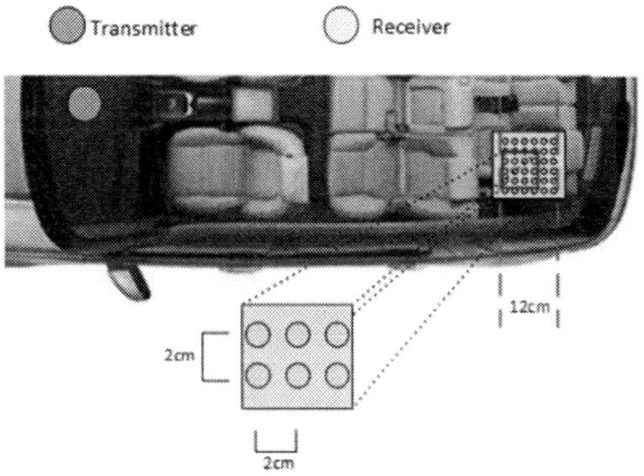

Fig. 2. Measurement grid of 12cm by 12cm (36 locations) for Test-2.

b) Test-2: Measuring the received signal power at 2.4 GHz in the boot compartment within a small 12cm x 12cm grid (36 locations with a separation distance of 2 cm between the marked locations), as shown in Fig. 2, for validating the loss component due to small-scale fading.

For each of the measurements taken, all necessary information such as the height and relative distance of the location was recorded and used later to calculate the separation distance between the transmitter and receiver antennas. The recorded results in the form of complex numbers were processed in MATLAB to extract the received power in dBm.

III. TEST RESULT AND ANALYSIS

A. Mean Path Loss

Path loss is frequently used to measure the propagation loss of radio signal transmission between a transmitter and a receiver. The intra-vehicle path loss behavior can be modelled by combining the mean path loss L_m, the loss due to slow or large-scale fading X_α, and loss due to fast or small-scale fading β_s. The combined path loss, denoted by L_p, is expressed by (all in dB):

$$L_p = L_m + X_\alpha + \beta_s \tag{1}$$

The mean path loss is defined by the generalized Friis equation also known as the log-distance path loss model, which comprises two parts: 1) the loss at a reference distance, d_0, from the transmitter and 2) the loss logarithmically increasing with distance, which is given by [7]:

$$L_m = L_{ref}(d_0) + 10nlog_{10}(\frac{d}{d_0}) \tag{2}$$

where the terms L_{ref}, n, d, d_0 represents the path loss at a reference distance, the path loss exponent, the distance between the transmitter and receiver antennas, and the chosen reference distance, respectively.

By using the measurements collected from the 44 locations in Test-1, the mean path loss L_m can be estimated by using the Least Square Linear Regression method [8]. Base on Eq.(2), the mean loss value for each of the 44 locations (receivers), $L_{m,i}$, is obtained by:

$$L_{m,i} = L_{ref,i} + 10n_i x_i \quad for\ i = 1, 2, ...44. \tag{3}$$

where

$$x_i = \log_{10}(\frac{d_i}{d_0}), \tag{4}$$

$$n_i = \frac{\sum_{i=1}^{44}(x_i - \bar{x})(L_i - \bar{L})}{\sum_{i=1}^{44}(x_i - \bar{x})^2}, \ and \tag{5}$$

$$L_{ref,i} = \bar{L} - n_i\bar{x} \ \ (\bar{L} = \frac{\sum_{i=1}^{44} L_i}{44}). \tag{6}$$

where d_i, L_i, represent the distance of ith location in meters and the instance path loss of the ith location in dB, respectively.

The path loss and estimated mean path loss which represent the first path loss component, L_m, are produced in Fig.3 for 2.4 GHz and Fig. 4 for 5.9 GHz. The reference distance is set to be $d_0 = 1$ meter. The extracted parameters for both 2.4 GHz and 5.9 GHz are shared and shown in Table II together with the loss at the reference distance, L_{ref}.

For comparison purposes, the results of the Friis model or the free-space loss model (L_f and $n = 2$) are also produced in Fig. 3 and Fig. 4, according to:

$$L_f = 20log_{10}(\frac{d}{d_0}) + 20log_{10}(f) + 20log_{10}\left(\frac{4\pi}{c}\right) \tag{7}$$

where f and c represent signal frequency and the speed of light, respectively. These results show that the mean path loss of the intra-vehicle narrowband transmission exhibits varied relationships with the free-space loss. At 2.4 GHz the mean loss is lower than the free-space loss by 2-4 dB over the 3 meters range in the vehicle, as shown in Fig. 3. However, when the frequency is increased to 5.9 GHz, the mean loss has a significantly larger increase than that of the free-space loss over the majority of the test range, as shown in 4. It is also observed that the special path loss performance inside vehicles, in terms of the scale of variation, is mainly contributed by the multipath fading effect which will be examined in detail in the following subsections.

TABLE II
ESTIMATED LOSS EXPONENT AND REFERENCE PATH LOSS

Frequency (GHz)	n	L_{ref} (dB)
2.4	2.212	37.04
5.9	1.289	50.43

TABLE III
GOODNESS OF FIT FOR LARGE-SCALE FADING (X_α)

Frequency: 2.4 GHz			
Distribution	KS	Chi	MLE
Log-normal	75.55%	23.25%	37.97%
Rayleigh	0.02%	5.44%	<0.01%
Rician	6.92%	23.85%	23.62%
Weibull	0.41%	23.67%	2.92%
Nakagami	17.10%	23.79%	35.5%
Frequency: 5.9 GHz			
Distribution	KS	Chi	MLE
Log-normal	91.17%	20.76%	65.06%
Rayleigh	0.1%	24.73%	<0.01%
Rician	1.92%	18.12%	7.92%
Weibull	<0.1%	16.98%	2.8%
Nakagami	6.72%	19.42%	24.22%

B. *Large-scale fading*

The large-scale fading also known as shadowing represents the local average slow fading characteristic of the received signal. It is defined in this analysis as the local-mean received signal power within a window of 50 cm which is equivalent to 4λ at 2.4 GHz and 10λ at 5.9 GHz, where λ denotes the wavelength. This window size is chosen base on the consideration that it should cover a enough number of measurements from neighboring locations in the test. But at the same time it should not be too wide as this will compromise the location differentiation factor, i.e. no obvious variation in loss can be identified between neighboring local mean values. The loss component due to large-scale fading is extrapolated using the Moving Mean method, with the metric given by [9]:

$$L_\gamma = \frac{1}{k}\sum_{i=1}^{k} L_i \tag{8}$$

where L_γ represents the local mean, k is the window size defined by the number of samples within the window, and L_i is the path loss of the ith sample. The large-scale fading results are shown in Fig. 3 for 2.4 GHz and Fig. 4 for 5.9 GHz, by the dashed lines.

In wireless communications, large-scale fading is normally described as a log-normal distributed random variable. But it has been argued that this is not always the case as other distribution functions have demonstrated a better fit in some scenarios. Therefore, we evaluated this performance using the Cumulative Density Function (CDF) of the Relative Loss Variation (RLV) against the mean path loss, and through the comparisons between different distribution functions such as

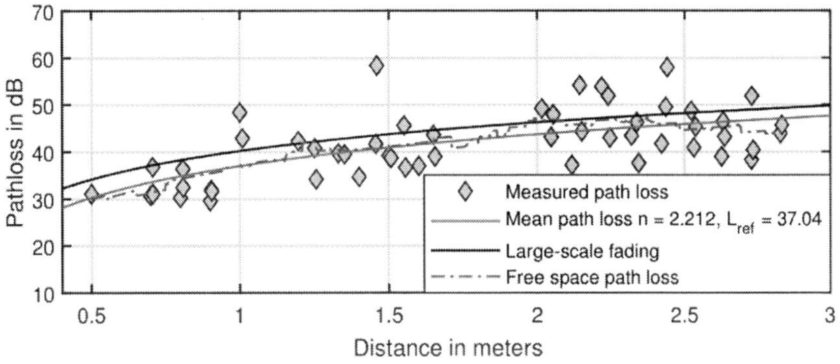

Fig. 3. Measured path loss, estimated mean path loss and loss due to large-scale fading at 2.4 GHz.

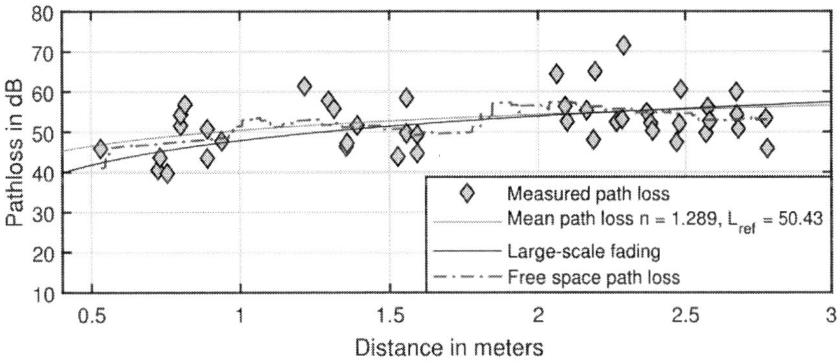

Fig. 4. Measured path loss, estimated mean path loss and loss due to large-scale fading at 5.9 GHz.

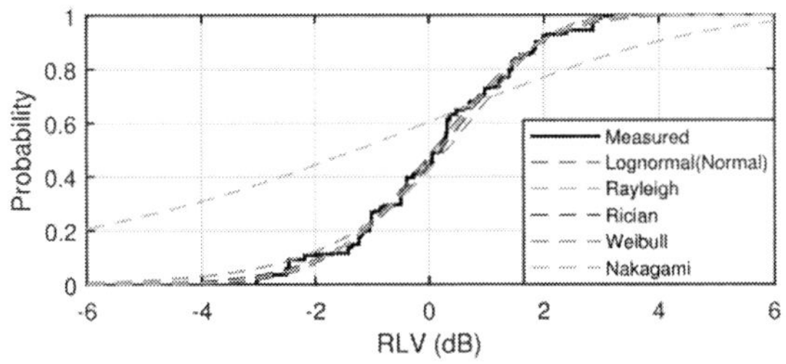

Fig. 5. CDF of relative loss variation for large-scale fading at 2.4 GHz.

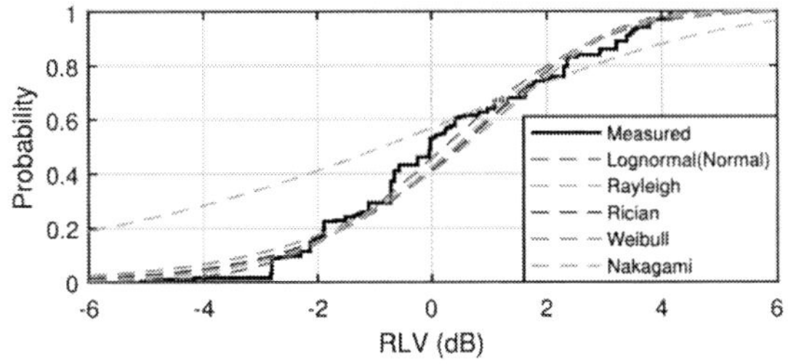

Fig. 6. CDF of relative loss variation for large-scale fading at 5.9 GHz.

TABLE IV
PARAMETERS OF LOG-NORMAL DISTRIBUTION (NORMAL IN DB) FOR LARGE-SCALE FADING (X_α) AT 2.4 GHZ AND 5.9 GHZ

Frequency(GHz)	Mean(dB)	Standard Deviation(dB)	Variance(dB)
2.4	0.0824	1.479	2.188
5.9	0.236	2.121	4.5

log-normal, Rayleigh, Rician, Nakagami and Weibull distributions, as shown in Figs. 5 for 2.4 GHz and 6 for 5.9 GHz.

To show the Goodness of Fit (GoT) of those distribution functions to our measurements, we applied three different test metrics: Kolmogorov–Smirnov (KS), Chi-Square (Chi) and Maximum Likelihood Estimation (MLE), with results tabulated in Table III. The KS results suggest the log-normal distribution being the highest match percentage with big margins from the others. However, the Chi and MLE results show no dominance of the log-normal distribution and, in particular, the Chi results have demonstrate that Rician, Nakagami and Weibull are all better fitted than log-normal at 2.4 GHz. The parameters that represents the large-scale fading X_a as log-normal (normal id dB) for both frequencies are shared in Table IV.

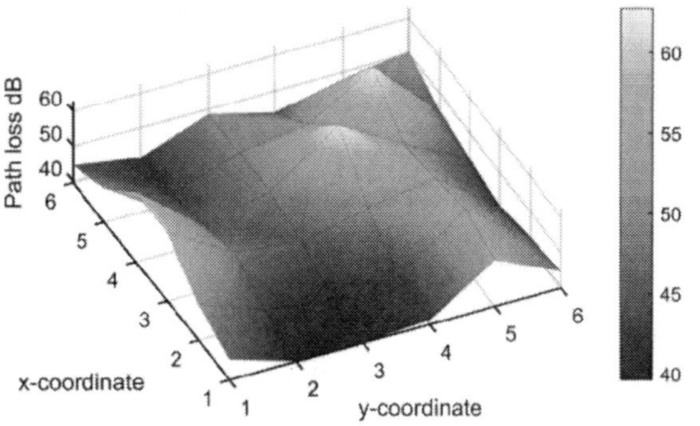

Fig. 7. A 3-dimensional view of the measured path loss in the grid at 2.4 GHz with 2 cm spacing.

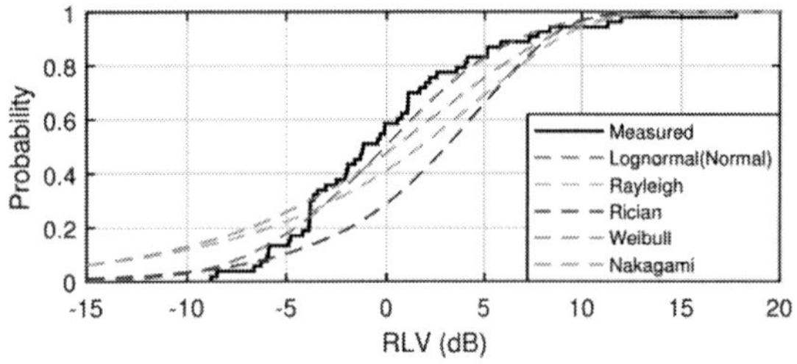

Fig. 8. CDF of relative loss variation for small-scale fading at 2.4 GHz.

C. Small-scale fading

Small-scale fading is defined as the variation of the received signal power over a short time period or a short distance less than a few wavelengths. The magnitude of this variation can be as high as 40 dB [10]. In our work, the intra-vehicle small-scale fading profile was investigated using two methods based on the measurements taken from a spatially tightened grid, as described below.

a) Spatial distribution of path loss due to small-scale fading: The received signal power was measured and later converted into the path loss for 36 locations as shown in Fig. 2. The distribution of the path loss for those location is shown by Fig. 7 in a three-dimensional view. The result shows a significant loss variation when the receiver is moved by just 2 cm. This behavior is due to the multipath characteristics of the channel and in particular due to the narrowband of signals transmitted. In addition, the largest variation in path loss measure at 55 dB, which is 15 dB above the mean loss of this grid, is observed.

b) CDF for small-scale fading effect: The small-scale fading component was extracted from the measured path loss from Test-1 as shown in Fig. 1, by deducting the large-scale fading component from the actual loss measurement which represents the relative loss variation (RLV) for small-scale fading. The small-scale fading effect was then demonstrated using the CDF of RLV in comparison with the same group of popular distribution functions, as shown in Fig. 8 for 2.4 GHz and Fig. 9 for 5.9 GHz.

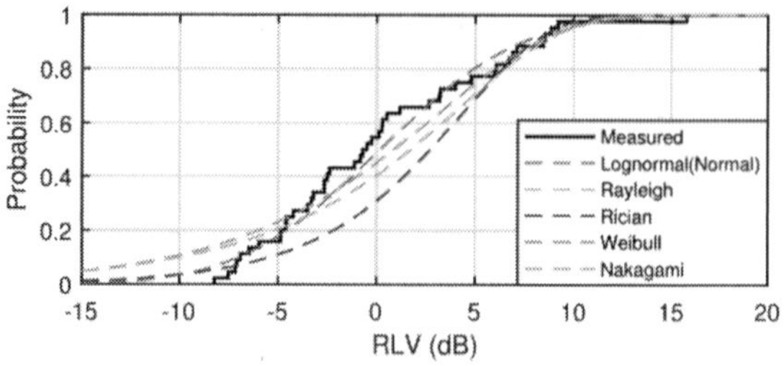

Fig. 9. CDF of relative loss variation for small-scale fading at 5.9 GHz.

TABLE V
PARAMETERS OF LOG-NORMAL DISTRIBUTION FOR SMALL-SCALE FADING (β_s) AT 2.4 GHZ AND 5.9 GHZ

Frequency(GHz)	Mean(dB)	Standard Deviation(dB)	Variance(dB)
2.4	-0.072	5.339	28.505
5.9	0.177	5.617	31.554

By comparing the CDFs in Figs. 5 and 6 with those in Figs. 8 and 9, clearly small-scale fading lays a considerably dominant role over large-scale fading in contributing to the overall path loss in the intra-vehicle environment. The RLV caused by small-scale fading is significantly higher than that of large-scale fading and can reach as high as 18 dB.

To check the GoT among those distributions to the measured results, KS, Chi and MSE metrics were also used with the comparison results shown in Table VI. As indicated by KS and MLE results, the log-normal distribution is statistically a better fit than other types of distributions. Some statistical parameters for describing small-scale fading β_s as a log-normal distribution (normal in dB) for both frequencies are given in Table V.

The small-scale fading distributions for both frequencies appear to be similar to each other. This feature has been confirmed by using the two-sample KS test which demonstrates a

TABLE VI
GOODNESS OF FIT FOR SMALL-SCALE FADING (β_s)

Frequency: 2.4 GHz			
Distribution	KS	Chi	MLE
Log-normal	84.41%	24.68%	92.19%
Rayleigh	0.09%	15.4%	0.29%
Rician	0.09%	15.4%	0.29%
Weibull	14.41%	23.33%	5.81%
Nakagami	1.012%	21.19%	1.42%
Frequency: 5.9 GHz			
Distribution	KS	Chi	MLE
Log-normal	70.43%	23.14%	68.81%
Rayleigh	0.01%	15.73%	3.73%
Rician	0.01%	15.73%	3.73%
Weibull	24.58%	23.39%	15.50%
Nakagami	4.96%	22.02%	8.22%

TABLE VII
SMALL-SCALE FADING (β_s) TWO-SAMPLE-KS TEST

KS p-value		
Frequency (GHz)	2.4(Estimated)	5.9(Estimated)
2.4(Estimated)	1	0.984
5.9(Estimated)	0.984	1

close correlation between the small-scale fading components β_s at 2.4 GHz and 5.9 GHz, as shown in Table VII. In other words, the path loss variation caused by small-scale fading is not significantly affected by the carrier frequency in this environment.

IV. CONCLUSION

The paper has presented the propagation characteristic of narrowband signals at 2.4 GHz and 5.9 GHz in the intra-vehicle wireless channels. Various channel parameters have been extracted from the received signal power measurements such as the distance-related attenuation loss or the mean loss, large-scale fading and small-scale fading statistics, which jointly contribute to the overall path loss. In addition, we have demonstrated that the small-scale fading effect is not statistically dependent on carrier frequency.

We have found that multipath fading has a significant impact on the path loss performance of narrowband signals compared to the attenuation-related loss which has a varied relationship with the free-space loss depending on the operating frequency chosen. We have also showed that small-scale fading is much more influential than large-scale fading on path loss in this environment. These findings will help to identify suitable technologies to mitigate the small-scale fading effect on data transmission in intra-vehicle wireless channels.

REFERENCES

[1] F. Arena and G. Pau, "An Overview of Vehicular Communications," Futur. Internet, vol. 11, no. 2, p. 27, 2019.

[2] J. Blumenstein et al., "In-vehicle channel measurement, characterization, and spatial consistency comparison of 3-11 GHz and 55-65 GHz frequency bands," IEEE Trans. Veh. Technol., vol. 66, no. 5, pp. 3526–3537, 2017.

[3] L. Cheng, J. Casazza, J. Grace, F. Bai, and D. D. Stancil, "Channel Propagation Measurement and Modeling for Vehicular In-Cabin Wi-Fi Networks," IEEE Trans. Antennas Propag., vol. 64, no. 12, pp. 5424–5435, 2016.

[4] R. Liu, S. Herbert, T. H. Loh, and I. J. Wassell, "A Study on Frequency Diversity for Intra-Vehicular Wireless Sensor Networks (WSNs)," in Proc. IEEE Veh. Technol. Conf. (VTC Fall), 2011, pp. 1–5.

[5] A. R. Moghimi, H. M. Tsai, C. U. Saraydar, and O. K. Tonguz, "Characterizing intra-car wireless channels," IEEE Trans. Veh. Technol., vol. 58, no. 9, pp. 5299–5305, 2009.

[6] H. Kamoda, S. Kitazawa, N. Kukutsu, K. Kobayashi, and T. Kumagai, "Microwave propagation channel modeling in a vehicle engine compartment," IEEE Trans. Veh. Technol., vol. 65, no. 9, pp. 6831–6841, 2016.

[7] T. S. Rappaport, Wireless Communications Principles and Practice, 2nd ed., Prentice Hall: Upper Saddle River, NJ, USA, 2002.

[8] A. C. Rencher and G. B. Schaalje, Linear Models in Statistics, 2nd ed., Wiley, 2008.

[9] D. de la Vega et al., "Generalization of the Lee method for the analysis of the signal variability," IEEE Trans. Veh. Technol., vol. 58, no. 2, pp. 506-516, 2009.

[10] A. Grami, "Chapter 12 - Wireless Communications," in Introduction to Digital Communications, 2015.

A Multi-Agent Approach to Simulate Autonomous Traffic with Games: How to Transform GTA-SA/SA-MP in Your Simulation Platform

Jônata N. Cirqueira*, Pedro C. Mesquita†, Rodrigo R. Novaes Jr.‡ and Sandro R. Dias§
Centro Federal de Educação Tecnológica de Minas Gerais
Belo Horizonte, Brazil
E-mail: {jonatanc0511*, pedrocesar1410†, rodrigo.novaes.jr‡}@gmail.com, sandrord@cefetmg.br§

***Abstract*—Urban mobility is among the main problems of the contemporary society. In this context, the vehicle automation technologies stand out in several aspects, such as reduction in accidents, congestion and emission of pollutants. Considering that, this work seeks to upgrade the simulation platform composed by Grand Theft Auto: San Andreas, and its modification San Andreas: Multiplayer (GTA-SA/SA-MP) that, according to the literature, is an appropriate environment for implementing autonomous vehicles networks. By using available structures in this environment, it is possible to perform three-dimensional simulations in many scales, ranging from a single village to an extensive intercity map. From the available structures in game, it was used a set of navigational nodes, which identify all the available routes for the vehicles in the roads. We used the navigational nodes to generate a weighted directed graph, to which we applied Dijkstra's and A* search algorithms. From that, it was observed that the vehicles were able to calculate and follow the best route from a source to a target node in any place of the map, obtaining a realistic environment to simulate and test solutions for the traffic, such as the autonomous intersection management protocol, which will be implemented in future.**

Keywords*–*Autonomous vehicles; multi-agent system; game; VANET; IoT.

I. INTRODUCTION

Urban mobility has proved to be one of the main problems we face in society. According to Texas A&M Transportation Institute, in 2017 the delays of industrial deliveries, along with the fuel consumption, caused a congestion cost of US$ 179 billion in the 494 urban areas of United States. Moreover, an average auto commuter wasted 21 gallons of fuel only at traffic congestion. The report also attested that average auto commuters spent 71 hours of extra travel time in the same year [1].

Besides monetary cost, many fatalities happen in the traffic. According to the World Bank Group, about 1.25 million people die on world's roads every year, while in average 20 or 50 million people are seriously injured. In America, traffic has been the leading cause of death since 1975 [2].

In addition, traffic congestion is a major aggravating factor for environmental problems. Vehicle emissions became the dominant source of air pollutants, raising the risks of morbidity and mortality, specially for commuters and individuals who live near roadways. When traffic flow is slow, regular speedups and slowdowns can increase travel time and diminish the dispersion of pollutants, elating by four times the emission of CO, HC and NO_x [3].

One of the emerging solutions for these problems is the traffic automation, currently leaded by the development of autonomous vehicles. They are proving to be effective and efficient, mainly due to their ability of taking deterministic and accurate decisions. Researches indicate that a vehicular network composed by 5% of self-driving cars already demonstrates significant advantages in the traffic flow [4].

Many simulation platforms were built to test autonomous traffic problems and solutions. Having a variety of them is important to validate different scenarios and perspectives. The two main perspectives for autonomous vehicles simulations are two-dimensional (2D) and three-dimensional (3D). The 3D perspectives are closer to reality, containing more details and problems to solve, like relief dynamics, consumption cost when driving uphills and downhills, among others [5]. However, due to the complexities and costs in building 3D software, most of the simulations are 2D, like the Texas University's AIM project [6][7].

Considering that, games like *Grand Theft Auto: San Andreas* (GTA-SA) [8] and its modification *San Andreas Multiplayer* (SA-MP) [9] team up to form a good environment to simulate multi-agent systems. This platform leverages many of the game's features, like an extensive 3D map containing multiple cities, physical models for mechanics and collision, different vehicle models, weather manipulation and many others [10].

Previous works have managed to successfully build an autonomous vehicles network inside GTA-SA/SA-MP [10]. Some of them could even build autonomous intersection management protocols, which is one of the hottest topics in autonomous traffic problems [11]. However, these works were expanded only to a small village in the game, being unable to scale all solutions for the whole game map [10]. Therefore, this work sought to implement an autonomous traffic system using the entire GTA-SA's map, whose detailed goals can be described as:

- Use the game's structures to identify vehicle's paths along the game map;
- Apply routing algorithms to understand if the game provides a consistent environment for simulations.

We could adapt GTA-SA/SA-MP's features to simulate autonomous traffic in the whole game map. Also, we found paths data for different objects, like pedestrians and boats, opening possibilities to interact with human agents and to develop sensor networks for sea navigation.

In Section I, the reader had a vision of the existing problems traffic automation want to solve, as well as some tools that can be used to achieve that. In Section II, we will present the base concepts and methodologies applicable. Section III describes the implemented methodology in a way to achieve the described goals. In Section IV, we expose the results and data acquired from this work. In Section V, we do a critical analysis on the data, as well as briefly explore future works.

II. Concepts & Methodology

This section will use theoretical concepts to present methods to achieve our goals. In Section II-A, we do an analysis on how to use games to simulate an autonomous multi-agent system; this will require software engineering and design patterns understanding. In Section II-B, we'll understand which math and computational models, as well as algorithms, are necessary to apply in this system. In Section II-C, we define the problems we want to solve with this model, as well as the high-level solution that might fit in any kind of system.

A. Games and design patterns

We can use two object-oriented programming principles to say that an electronic game or software product is eligible to simulate autonomous agents and multi-agent systems: open/close and the dependency inversion principles [12].

Definition 1: **Eligibility of a software product to be used in modifications.** Let M be a software product, M can only be extensible to modifications when the following characteristics are present:

1) M's modules and libraries may be extended without being modified;
2) There is a set of public interfaces of M such that they may be implemented, incorporated and distributed with M.

By identifying an application that fits Definition 1, we will reduce and decouple the amount of work. The responsibility of simulating autonomous agents and multi-agent systems can be addressed to the researcher, while the dependencies of this work, like physical models, climate, objects and events can rely on the modified application only.

B. Routing & mechanics models

Graph theory is the study of the relation between elements of a set. $G = (V, E)$ is said to be a graph, where V is a set of vertexes and $E = (u, v)$ is a set of pairs, where $u, v \in V(G)$, meaning that a vertex u is related to v. A graph is directed when there is a relation over all edges $(u, v) \in E(G)$ such that u is said to be incident over v [13].

Graphs can be used to understand whether elements of a set are reachable from a source vertex. Also, we can apply the concept of paths or routes, where there is a sub-graph $P = (V, E)$ such that $P \subseteq G$, whose $V(P) = \{u_1, u_2, \ldots, u_k\}$, $k \leq |V(G)|$, where all edges are in the format $E(P) = \{u_1u_2, u_2u_3, \ldots, u_{k-1}u_k\}$. Basically, a route or a path is a sub-graph that establish a traversal inside the original graph [13][14].

Also, if there is a cost function $C : E(G) \rightarrow \mathbb{R}$, where $\mathbb{R}$ is the set of real numbers, then we can calculate a route $P = (V, E)$ such that the cost to traverse $E(P)$ is said to be the distance of P. For instance, we say that the distance between two vertexes is the sum of all costs associated to the edges that connect them [14].

Lemma 1: **A path with minimal cost.** Let $G = (V, E)$ be a graph and $C : E \leftarrow \mathbb{R}$ a cost function, a path $P \subseteq G$ is said to be minimal if all sub-paths in P are also minimal.

Proof: Suppose that $P = \{P_1, P_2, \ldots, P_k, \ldots, P_n\}$, where P_1 is a sub-path of P that passes through P_k and reaches all the way to P_n. Now, P is minimal if all $P_i \subset P$ are also minimal.

Let's prove it by contradiction. Suppose that P_k is not minimal, then we know that $P = \{P_1, \ldots, P_{k-1}, P_k, P_{k+1}, \ldots, P_n\}$ have sub-paths from P_1 to P_{k-1} and P_{k+1} to P_n which are minimal. However, there is a minimal path P'_k whose distance is lesser than P_k. Therefore, the distance of $P' = \{P_1, \ldots, P_{k-1}, P'_k, P_{k+1}, \ldots, P_n\}$ is lesser than P, meaning that concatenating minimal sub-paths will result in a minimal path. ■

The process of identifying a path with minimal cost is based on Lemma 1. Therefore, Figure 1 lists an algorithm to calculate all minimal paths from a single-source vertex of a graph.

for all $u \in V(G)$ **do**
  $d(u) \leftarrow \infty$ {Initial distance is unknown}
  $p(u) \leftarrow \emptyset$ {Identify which vertex is incident over u in the minimal path}
end for
$d(s) \leftarrow 0$
INSERT_HEAP(A, s) {Creates a priority queue A based on the distance to s}
while $|A| \geq 1$ **do**
  $u \leftarrow$ REMOVE_HEAP(A) {Removes the vertex with minimal distance from A}
  for all $uv \in E(G), v \in V(G)$ **do**
    if $d(v) > d(u)+$ CALCULATE_COST(uv) **then**
      $d(v) \leftarrow d(u)+$ CALCULATE_COST(uv)
      $p(v) \leftarrow \{u\}$
      INSERT_HEAP(A, v)
    end if
  end for
end while
return (p, d) {Returns a pair with the minimal distances d to all vertexes and a set p that identifies incidences}

Figure 1. Algorithm to calculate the minimal cost of a path inside a graph $G = (V, E)$, such that the path starts from $s \in V(G)$ and terminates in all other reachable vertexes in V.

The algorithm in Figure 1 can be implemented in two well-known versions: Dijkstra's algorithm for shortest path and A* algorithm [15]–[17]. The difference between them is how the operation CALCULATE_COST is implemented. Please look at Figure 2 for each version.

return $C(u, v)$

(a) Standard cost function for Dijkstra's algorithm for shortest path.

return $C(u, v) + h(v)$

(b) Standard cost function for A*'s algorithm for shortest path.

Figure 2. Implementations for each cost function according to Dijkstra's and A* algorithms in a graph $G = (V, E)$, where $u, v \in V(G)$ are two vertexes, $C : E(G) \rightarrow \mathbb{R}$ is a cost function associated to the edge $uv \in E(G)$ and $h(v)$ is any consistent heuristic.

One of the best heuristics is the euclidean distance between two points in the space. Considering a map of three-dimensional coordinates, the distance between two points $P_1 = (x_1, y_1, z_1)$ and $P_2 = (x_2, y_2, z_2)$ is given by:

$$d(P_1, P_2) = \sqrt{(x_2 - x_1)^2 + (y_2 - y_1)^2 + (z_2 - z_1)^2}, \quad (1)$$

which can be applied as the cost function of a graph when its vertexes represent coordinates of a 3D map.

C. Problems and definitions

Since routing problems are modelled using graph theory, one way of modelling and solving traffic problems is combining graph theory with a multi-agent approach [13]. In this context, Problems 1 and 2 present the definition of an autonomous traffic model, whose agents are named as autonomous vehicles.

Problem 1: **Model an autonomous traffic system.** Given a set of vehicles A and a set of paths $P = (V, E)$, where V is a set of positions and $E = (u, v)$, $u, v \in V$, is a set of pairs that indicate that v is reachable through u, output a set of routes $R_i = (V_i, E_i)$, $\forall R_i \subseteq P$ that a vehicle a_i, $\forall a_i \in A$, can traverse autonomously.

Problem 2: **Model an autonomous vehicle.** Given a vehicle a and a set of paths $P = (V, E)$, where V is a set of positions and $E = (u, v)$, $u, v \in V$, is a set of pairs that indicate that v is reachable through u, output route $R = (V', E')$, $R \subseteq P$ such that a can traverse autonomously.

According to the definitions of Problem 1, considering G a set of graphs and V a set of vertexes, let $\Phi : G \times V \to G$ be a function that receives $P = (V, E)$ and an initial vertex $s \in V(P)$, such that it outputs a path $R \subseteq P$. In essence, Φ outputs a route starting from vertex s. Also, let $\Omega : V \to V$ be a function that receives a set of vertexes and outputs a random vertex of the set. Henceforth, consider the algorithms in Figure 3 as solutions for Problems 1 and 2.

$R \leftarrow \emptyset$
for all $a \in A$ **do**
  $R_a \leftarrow \emptyset$
  $s \leftarrow \Omega(V(P))$
  $R_a \leftarrow R_a \cup \Phi(P, s)$
end for
return R

(a) Solution proposed for Problem 1.

$R \leftarrow \emptyset$
$s \leftarrow \Omega(V(P))$
$R \leftarrow R \cup \Phi(P, s)$
return R

(b) Solution proposed for Problem 2.

Figure 3. Solutions for the problems proposed in the paper. Consider R a set of routes, R_a a subset of routes for a vehicle a, $s \in V(G)$ a source vertex for a graph P, where $P = (V, E)$. $\Phi : P \times V \to P$ is an operator that returns a path given a graph and a source vertex.

According to the algorithms in Figure 3, Φ can be any single-source routing algorithm in a directed graph. Some examples are breadth-first and depth-first searches, which detect all reachable vertexes from a single source [13]. There are also shortest path algorithms, such as Dijkstra's and Bellman-Ford's, which calculate the minimal paths from a single source to every reachable vertex of the same graph [15][18]. Heuristics, such as A*, are welcomed as well [16][17]. Notice that the latter algorithms require a cost function $C : E(G) \to \mathbb{R}$, where $\mathbb{R}$ is the set of real numbers.

Another important topic is reducing large amounts of work into sub-problems, which allows us to avoid repetition and implement reusable design patterns [14]. Lemma 2 applies this vision to Problems 1 and 2.

Lemma 2: The problem of modelling an autonomous vehicle is a sub-problem of modelling an autonomous traffic system.

Proof: Let $A = \{a_1, a_2, \ldots, a_n\}$ be a set of n vehicles and $P = (V, E)$ a graph, where V is a set of positions and $E = (u, v)$, $u, v \in V$ a set of pairs that indicate that v is reachable through u. Consider the following assumptions:

- There is a set of known, but randomly calculated source vertexes $S = \{s_1, s_2, \ldots, s_n\}$, where $S \subseteq V$;
- The solutions of Problems 1 and 2 will always calculate the route for a vehicle a_i starting from the corresponding vertex s_i, where $s_i \in S$;
- The algorithm used to calculate the route R_i for a_i will be the same in both solutions.

Given a set of routes $R = \{R_1, R_2, \ldots, R_n\}$ calculated by the solution of Problem 1, we define R' a set of routes that will be calculated individually by all vehicles $a \in A$, as said by the definition of Problem 2. According to Lemma 2, if $R' = R$ we can prove that Problem 2 is a sub-instance of Problem 1.

Let's prove it by induction. The steps are:

1) Initially, $R' = \emptyset$
2) For all vehicles $a_i \in A$, apply the solution of Problem 2, generating a route R'_i;
3) Insert the solution R'_i to R';
4) R'_i was generated with the same algorithm as R_i and starting from the position s_i, therefore $R'_i = R_i$.

Since every $R'_i \in R'$ is equals to the corresponding $R_i \in R$, then the sets R' and R are equal. Therefore, we proved that applying the solution of Problem 2 to all vehicles in A will produce a valid solution for Problem 1, hence Problem 2 is a sub-instance of Problem 1. ■

Lemma 2 is useful because it tells us that, to model an autonomous traffic system, we only need to model a set of autonomous vehicles. Autonomous traffic is hence a composition of autonomous vehicles.

III. Development

This section will present how we applied the methodology in our work. In Section III-A, we describe the development environment and how we used it to simulate a multi-agent system. In Section III-B, we do a more detailed analysis on how to use the environment's tools to model the agents. Then, in Section III-C we describe which algorithms and data structures we designed in order to make the environment behave like an autonomous vehicular traffic system.

A. GTA-SA/SA-MP: an online multiplayer game server

The GTA-SA/SA-MP is a multiplayer modification for the Rockstar's game GTA-SA [8][9]. It uses the dependency inversion principle to make an interface between an abstract machine interpreter and the game itself [12].

The abstract machine, AMX, is the compilation result of a well-known scripting language, named Pawn [19]. SA-MP is an AMX interpreter that executes the compiled AMXs with the developers' modifications. Therefore, GTA-SA/SA-MP is a client-server application, where GTA-SA is the client and SA-MP is the server [9].

Also, Pawn interpreters follow a design pattern that provides a common interface to be extended in separated plug-ins. For instance, developers can write dynamic libraries that include the AMXs' interpreters base functions, allowing them to be extended and reused in Pawn scripts. Since dynamic libraries are developed in middle-level languages, such as C or C++, we can improve our server for better memory management and computational cost [9][12][19].

When a player first enters a SA-MP server, it will establish a connection between the game and the respective server, whose events, objects and other dynamics will be controlled and modified according to the compiled AMXs [9][20].

In this context, we created a set of Pawn scripts and dynamic libraries that used the game's built-in objects, physics and events to modify its default behavior, allowing us to introduce our own functions to simulate an autonomous vehicle system.

SA-MP has a consolidated community that writes and shares plug-ins and scripts around the world [20]. We used some of these, like the *Fully Controllable Non-Playable Character* plug-in, FCNPC, which allowed us to insert and control non-playable characters, NPCs, in the game map [21]. This plug-in was important to implement the agent system's behaviors, which can be seen in more details in Section III-B.

Also, based on the premise that games render only a limited amount of objects in the screen, mostly to save memory and computational resources, it was important to use plug-ins that embodied this principle and avoided extra memory consumption. Therefore, we used *SA-MP Streamer Plugin* to draw objects and labels in the game map, allowing us to display some structural information about the system [22].

In addition, SA-MP exports a set of paths' data that we could decode and use to implement our system [23]. Please follow to Section III-C to have more details on how we modified the game's path structures to implement routing algorithms.

Finally, we used the described tools and data structures to solve Problem 2, which can be scaled to a solution of Problem 1.

B. Autonomous agents modelling

GTA-SA/SA-MP provides a whole set of objects and structures that can be modified by FCNPC plug-in [20][21]. Among them, there are non-playable characters and vehicles. We used these elements to solve Problems 1 and 2, proposed in this work's methodology at Section II-C.

In this case, vehicles can be controlled either by the player or by an NPC. Since NPCs can have autonomous behavior, then we could simulate autonomous driving by inserting a NPC into a vehicle [21]. Therefore, we created a dynamic library that extends SA-MP server to do basic operations over NPCs and vehicles, given by Definition 2.

Definition 2: **The Driver NPC plug-in**. It is a dynamic library that implements SA-MP's model to manage NPCs directly inside a vehicle. This plug-in exports three functions:

- *Create*: receives as input the vehicle's attributes, as a pair of primary and secondary colors C_1 and C_2, a vehicle model T, an initial position s and an angle θ in relation to the north, which indicates the direction the vehicle will be facing. It renders the model in the described position and will return a unique identifier a, representing the NPC;
- *Destroy*: receives as input an NPC identifier a. It will destroy both NPC and its corresponding vehicle from the screen;
- *Move*: receives as input an NPC identifier a and a pair of positions u and v. It will make the NPC travel from u to v simulating an autonomous driving.

In Definition 2, the operations *create* and *destroy* were developed using GTA-SA/SA-MP's own features [20][24]. However, operation *move* needed to be adapted in different situations, like curves and hills. In curves, we calculated the Bezier's curve between the initial and final positions [25]. Driving up and down hills was still an obstacle, and we couldn't find a method to adjust the vehicle's angle while traversing different reliefs.

C. Routing algorithms and data structures

GTA-SA/SA-MP provided a set of path structures that allowed us to model positions and hence a directed graph in the game map [23]. The structures are described in Definition 3.

Definition 3: **SA-MP's links and nodes**. There are two main types of nodes exported by GTA-SA/SA-MP:

- *Path-nodes*: a structure containing an identifier i, an area identifier p, a link identifier l and a position $P = (x, y, z)$ in the game's 3D space. Path-nodes are placed in streets and roads along the map. Also, they can be extended in different kind of nodes:
  - *Navi-nodes*: a structure containing detailed information about a path-node. It has the referred path-node's identifier i and area identifier p, and a position $P = (x, y, z)$ in the game's 3D space, usually between two adjacent path-nodes. It also has a set of flags, like the amount of left and right lanes in relation to the navi-node and a value to indicates if the traffic flow is allowed in the right or left lanes;
  - *Ped-nodes*: describe the game's default paths for pedestrians. Usually placed on streets intersections, houses and sidewalks;
  - *Vehicle-nodes*: describe the game's default paths for vehicles. Usually placed on streets, roads and parking lots;
  - *Boat-nodes*: describe the game's default paths for boats. Usually placed at the sea.
- *Links*: a structure containing an identifier l and pairs of nodes and area identifiers $(i_1, p_1), (i_2, p_2)$, indicating that the nodes i_1 and i_2 in areas p_1 and p_2 have a connection.

Therefore, we created a dynamic library that extends SA-MP server to do operations over map positions, which allowed us to implement the proposed routing model. Hence, path-nodes and all their derivations could be inputted as positions; the *move* function, exported by the Driver NPC plug-in, as described in Section III-B, received a pair of path-nodes which represent the initial and final positions of the autonomous vehicle movement.

Definition 4: **The Paths plug-in**. It is a dynamic library that implements SA-MP's model to manage paths and positions in GTA-SA. It exports the following functions:

- *GetRelativeStreetPosition*: receives as input the identifiers i_1 and i_2 of source and target path-nodes and the vehicle's width w, returning a position $P = (x, y, z)$ in the correct lane for the vehicle to traverse;

- *Dijkstra*: receives a pair of random path-node identifiers s and t as input, returning a list of path-nodes identifiers, representing the route with minimal cost between path-nodes s and t calculated by Dijkstra's algorithm [15];
- *AStar*: receives a pair of random path-node identifiers s and t as input, returning a list of path-nodes identifiers, representing the route with minimal cost between path-nodes s and t calculated by A*'s algorithm [16].

The Definition 4 states that the Paths plug-in exports a set of functions related to routing calculation and path-nodes manipulation. The function *GetRelativeStreetPosition* needs a pair of source-target nodes because, according to Definition 3, path-nodes identify a street rather than a specific lane. Therefore, we used both nodes' angles to identify in which track of circulation the vehicle was and, as a consequence, in which lane it should be placed. For that purpose, we also used the vehicle's width to calculate the agent's correct position in the lane, $P : A \to \mathbb{R}$, given by

$$P(a) = c + \frac{w(a)}{2}, \qquad (2)$$

where $a \in A$ is a vehicle, $w : A \to \mathbb{R}$ is a function that returns the vehicle's width and c is a constant factor.

Also, this function was developed in a way that allows implementing both traffic ways, meaning that the agents can traverse in the positive lane direction as well as in the negative lane direction.

Finally, we developed the functions *Dijkstra* and *AStar* to calculate routes using Dijkstra's and A*'s algorithms, respectively, both returning a list of path-nodes, where the cost of navigating between a pair of path-nodes is given by the distance between two points in a 3D space, as defined in the equation (1) [15][16].

IV. RESULTS

This section will explore the results we obtained after applying the proposed methodology to GTA-SA/SA-MP. We indicate how many path structures were found in the game, as well as their category according to Section III-A. Then, we clarify how they were used to generate routes in order to simulate the autonomous traffic. Finally, we raised time execution metrics to understand if the generated routes match the expected cost for the applied algorithms.

In our development, we were able to identify a high number of path-nodes in the game, as displayed by Table I. According to it, the pedestrian nodes represent the majority of GTA-SA paths data, being followed by vehicles and boats.

TABLE I. QUANTITY OF PATH-NODES DISCOVERED IN THE GAME, GROUPED BY THEIR TYPES.

Type	Count
Pedestrian	37,650
Vehicles	30,587
Boats	1,596
Total	69,833

From Table I, we can also estimate that GTA-SA's default features allow an autonomous agent network to be composed by 54% of humans, 44% of vehicles and 2% of boats, considering the default path-nodes distribution.

In addition, the path-nodes and navi-nodes were processed into a graph $\Gamma = (I, L)$, where I is a set of path-nodes and L is a set of links, containing information regarding the connection between the nodes in the game map. Figures 4 and 5 show a visual representation of path and navi-nodes, respectively.

In Figure 4, we see the detailed information of path-nodes, like the identifier $i = 36$, the area identifier $p = 261$, the three-dimension position $P = (x, y, z)$ and a set of links $L = \{(36, 262), (36, 265), (36, 270)\}$, where each pair (i', p') represents a target path-node's identifier and area identifier, respectively.

Figure 4. Visual representation of a path-node and its attributes.

In Figure 5 we see detailed information about navi-nodes. They are mostly present on hills, curves and multi-lane streets or roads. We used their target path-node identifiers $i = 37$ and area identifier $p = 275$, as well as the amount of left and right lanes, which in the example counts as one for both.

Figure 5. Visual representation of a navi-node and its attributes.

Furthermore, we could categorize two different street models in the game: the single-lane and the multi-lane. The single-lane model has a single traffic way available or, more specifically, a way where navi-nodes expose zero left and right lanes. In the multi-lane model, there might be one or more traffic ways available for which the navi-nodes expose more than zero left and right lanes.

According to Figure 6, in the single-lane model the vehicle a can traverse in both traffic ways to reach a path-node i. This can be done by inverting all links' directions.

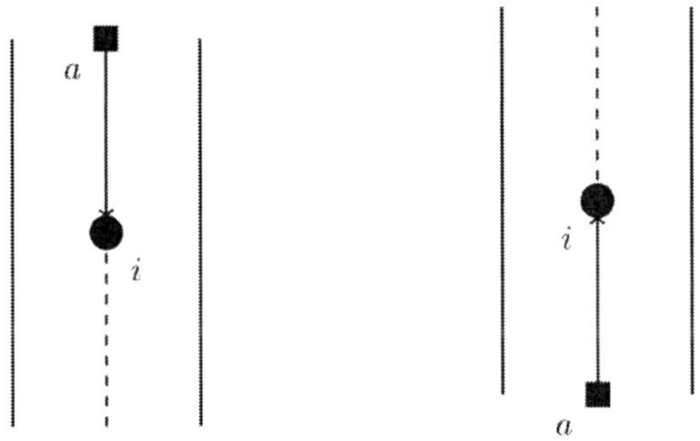

(a) North-south traversal in a single-lane street.

(b) South-north traversal in a single-lane street.

Figure 6. A single-lane street on which a vehicle a is reaching a path-node i.

According to Figure 7, in the multi-lane model vehicles a_1 and a_2 can traverse both in the same street to reach path-node i, but in different lanes. Therefore, we applied the equation in (2) to determine the correct vehicle's positions. Since we used the same vehicle models, w is a constant; hence, from empirical approximations, we determined $w = 0.75$ and $c = 1$.

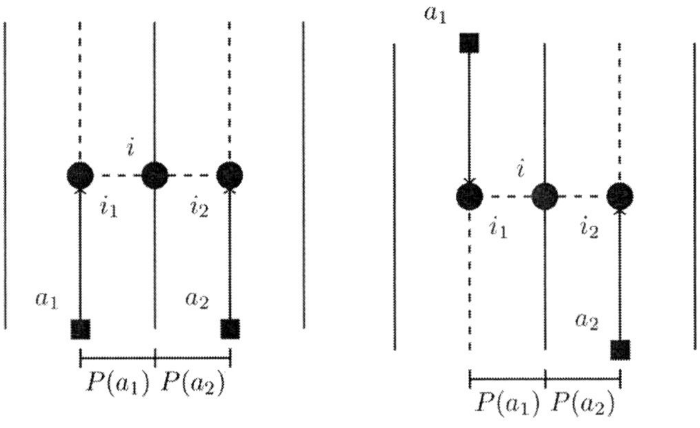

(a) Multi-lane street model with a single direction.

(b) Multi-lane street model with different directions.

Figure 7. Multi-lane street models. Vehicles a_1 and a_2 need to reach path-node i, but it's placed in the middle of the street. Therefore, we calculate the displacements $P(a_1)$ and $P(a_2)$ to determine the correct next positions i_1 and i_2, respectively.

As for the routes calculations, we recorded the time spent, in milliseconds, for both Dijkstra's and A* algorithms according to the number of path-nodes that composed the routes. The results were compiled in the chart on Figure 8. According to the chart, we noticed that A* is faster when generating routes for the same set of path-nodes. In addition, both algorithms calculated the same routes with minimal costs, meaning that the vehicle's path did not take any influence in the route computation.

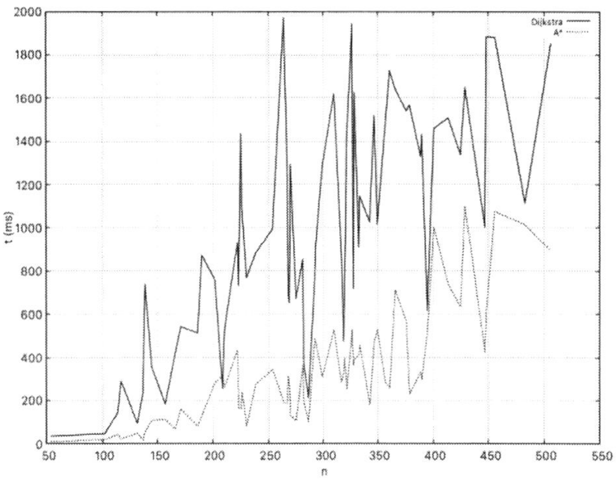

Figure 8. Time execution for the routing algorithms, in milliseconds (vertical axis), according to the number n of path-nodes (horizontal axis). The continuous line represents the time spent by Dijkstra's algorithm for shortest path, while the dotted one represents the time spent by A*'s algorithm.

The literature indicates that A* is faster than Dijkstra's algorithm whenever the heuristic is said to be consistent, which is also a good indicator that the adopted 3D space is consistent for routing models [16]. Also, we recorded a video of some simulations, which can be appreciated by the readers as a reference in this paper [26].

V. CONCLUSION

The results obtained in this work allowed a better simulation in the GTA-SA/SA-MP environment. When the autonomous vehicle network was extended to the whole map, it was possible to simulate different scenarios, taking advantage of many things that the environment offers, like different reliefs, street lengths and curve angles.

Furthermore, the results showed consistency with the reality, since A* and Dijkstra algorithms returned routes with minimal costs given the same set of path-nodes. Also, A* could calculate all routes with the proposed heuristic, which means that GTA-SA/SA-MP has a valid model of geometry and position.

This extension also allows the pedestrian and boat nodes to be implemented. Using the pedestrian nodes, it is possible to create an integration between autonomous vehicles and pedestrians in the roads, which is one of the main problems in autonomous vehicles systems. Using boat nodes, it is possible to simulate a maritime traffic, allowing to test the vehicles at sea as well.

Also, Autonomous Intersection Management protocols (AIM) can be implemented in a larger scale. Intersections are one of the main problems in autonomous vehicles systems too, since most of the traffic accidents happens in them. Therefore, implementing AIM at multiple intersections can be an efficient way to raise results about reductions in accidents and others factors, such as time spent, amount of emitted gases and spent fuel.

Another option would be increase the dynamics of the simulations, including weather manipulation, acceleration and deceleration in curves, uphills and downhills, reverse driving and overtaking. This would allow us to explore more scenarios, as well as increase the difficulty of the problems we are solving.

Now, beyond the successful simulations that were done in the whole GTA-SA map, the progress made in this work opens doors to perform simulations of the main problems of autonomous vehicles systems in a larger scale, using the realistic environment GTA-SA/SA-MP.

ACKNOWLEDGMENT

A very special thanks to our institution, Centro Federal de Educação Tecnológica de Minas Gerais, for all support given during this work. Also, to Daniel de Sousa Santos, for bringing this wonderful environment to our team and allowing us to work and have fun.

REFERENCES

[1] D. Schrank, B. Eisele, and T. Lomax, "2019 Urban Mobility Report," Texas A&M Transportation Institute, 2019, URL: https://static.tti.tamu.edu/tti.tamu.edu/documents/mobility-report-2019.pdf [accessed: 2020-03-17].

[2] W. Bank, "The High Toll of Traffic Injuries: Unacceptable and Preventable," World Bank Group Transport, 2017, URL: https://openknowledge.worldbank.org/handle/10986/29129 [accessed: 2020-03-17].

[3] K. Zhanga and S. Batterman, "Air pollution and health risks due to vehicle traffic," Science of The Total Environment, vol. 450-451, 2013, pp. 307–316, ISSN: 0048-9697.

[4] R. E. Stern, S. Cui, M. L. D. Monache, R. Bhadani, M. Bunting, and M. Churchill et al., "Dissipation of stop-and-go waves via control of autonomous vehicles," Transportation Research Part C: Emerging Technologies, vol. 89, 2018, pp. 205–221, URL: https://www.sciencedirect.com/science/article/pii/S0968090X18301517.

[5] S. Park, H. Rakha, K. Ahn, and K. Moran, "Fuel economy impacts of manual, conventional cruise control, and predictive eco-cruise control driving," International Journal of Transportation Science and Technology, 2013, pp. p. 227–242.

[6] T.-C. Au, S. Zhang, and P. Stone, "AIM: Autonomous Intersection Management for Semi-Autonomous Vehicles," Handbook of Transportation, Routledge, Taylor & Francis Group, 2015.

[7] "AIM: Autonomous Intersection Management," 2006, URL: https://www.cs.utexas.edu/~aim/ [accessed: 2020-03-17].

[8] Rockstar, "Grand Theft Auto: San Andreas," 2004, URL: https://www.rockstargames.com/games/info/sanandreas [accessed: 2020-03-17].

[9] "San Andreas: Multiplayer," 2006, URL: https://www.sa-mp.com/ [accessed: 2020-03-17].

[10] R. R. Novaes Jr., "Um Novo Ambiente de Simulação para Sistemas de Gerenciamento de Tráfego para Veículos Autônomos," Centro Federal de Educação Tecnológica de Minas Gerais, Tech. Rep., 2017.

[11] R. R. Novaes Jr., D. S. Santos, G. M. F. Santiago, and S. R. Dias, "A New Solution to the Traffic Managing System for Autonomous Vehicles (Demonstration)," in 16th International Conference on Autonomous Agents and Multi Agent Systems, São Paulo/SP, Brasil, May 2017, pp. 1805–1807.

[12] R. C. Martin, "Design Principles and Design Patterns," 2000, URL: https://fi.ort.edu.uy/innovaportal/file/2032/1/design_principles.pdf [accessed: 2020-03-01].

[13] R. Diestel, Graph Theory. Springer-Verlag, 2000, URL: http://www.esi2.us.es/~mbilbao/pdffiles/DiestelGT.pdf [accessed: 2020-03-01].

[14] T. H. Cormen, C. E. Lelserson, and R. L. Rilvest, Introduction to Algorithms, 3rd ed. MIT Press, 2009.

[15] E. W. Dijkstra, "A Note on Two Problems in Connexion with Graphs," Numerische Mathematik, vol. 1, Jun. 1959, pp. 269–271.

[16] W. Zeng and R. L. Church, "Finding shortest paths on real road networks," 2009, URL: http://doi.org/10.1080/13658810801949850 [accessed in 2020-03-01].

[17] A. Botea, M. Müller, and J. Schaeffer, "Near optimal hierarchical path-finding." J. Game Dev., vol. 1, no. 1, 2004, pp. 1–30.

[18] Walden and David, "The Bellman-Ford Algorithm and "Distributed Bellman-Ford"," Jan. 2008.

[19] CompuPhase, "Pawn Implementer's Guide," 2016, URL: https://github.com/compuphase/pawn/blob/master/doc/Pawn_Implementer_Guide.pdf [accessed: 2020-03-17].

[20] "SA-MP Wiki," 2017, URL: https://wiki.sa-mp.com/ [accessed: 2020-03-17].

[21] S. Marochkin, "SA-MP FCNPC Plugin," 2019, URL: https://github.com/ziggi/FCNPC [accessed: 2020-03-17].

[22] "SA-MP Streamer Plugin," 2014, URL: https://github.com/samp-incognito/samp-streamer-plugin [accessed: 2020-03-17].

[23] "Paths (GTA-SA)," SA-MP Wiki, URL: https://gta.fandom.com/wiki/Paths_(GTA_SA) [accessed: 2020-03-17].

[24] "Vehicle Models," SA-MP Wiki, 2019, URL: https://wiki.sa-mp.com/wiki/Vehicle_Models [accessed: 2020-03-17].

[25] J. London, New York, Ed., Numerical Control: Mathematics and Applications. J. Wiley, 1972, ISBN: 0471071951 9780471071952.

[26] R. R. Novaes Jr., P. C. Mesquita, and J. N. Cirqueira, "Autonomous traffic simulation with gta-sa/sa-mp," 2020, URL: https://bit.ly/3bfh0LN [accessed: 2020-03-20].

Implementation and Evaluation of Priority Processing by Controlling Transmission Interval Considering Traffic Environment in a Dynamic Map

Kohei Hosono*, Akihiko Maki†, Yosuke Watanabe‡, Hiroaki Takada‡ and Kenya Sato§
*Computer and Information Science, Graduate School of Science and Engineering, Doshisha University
Kyoto, Japan 610-0321
Email: kohei.hosono@nislab.doshisha.ac.jp
†Fujitsu Limited
Kanagawa, Japan 215-8588
‡Institutes of Innovation for Future Sciety, Nagoya University
Nagoya, Japan 464-8601
§Mobility Research Center, Doshisha University
Kyoto, Japan 610-0321

***Abstract*—Much attention has been attracted to the research of cooperative automatic driving that focuses on safety and efficiency by sharing the data obtained from sensor information of a vehicle. In addition, dynamic maps, a common information and communication platform for the integrated management of shared sensor information, are under consideration. A vehicle always sends data to a server that manages the dynamic map, and the server runs applications for driving support and control on the basis of the data, so fast information processing is required. However, congestion is a concern when data is continuously sent from vehicles to the server at high transmission intervals and when many vehicles are managed by dynamic maps on the server. In addition, the data transmission interval from the vehicle required by the road characteristics differs in actual traffic environments. Therefore, congestion can be alleviated by adjusting the transmission interval of data from the vehicle in consideration of road characteristics. In this paper, a platform for a dynamic map consisting of a server and a vehicle is constructed. We have also implemented a priority processing function that sets the priority for each section of a lane, and adjusts the transmission interval on the basis of the characteristics of the road around the vehicle.**

Keywords–ITS; Dynamic Map; Connected Vehicle; Automated Driving; Priority Processing; Load Balancing.

I. Introduction

In recent years, there has been a lot of research and development on automatic driving, where automobiles use sensors to recognize the surrounding environment and automatically control driving by avoiding hazards [1]–[3]. However, in-vehicle sensors are limited to detecting objects in the visible range but not in the inaccessible range. Therefore, cooperative intelligent transport systems (ITSs), which aims to improve safety by using wireless communication technology to exchange information between vehicles and roadside equipment, have attracted attention [4]–[6]. A variety of applications are being considered, including collision warning at intersections, provision of traffic jam and signal information, and support for merging on expressways [7]–[9]. However, the data sent from the vehicle is currently managed separately for each application. Therefore, dynamic maps, which are platforms for managing and processing data in an integrated manner, have been investigated [10]–[12].

A dynamic map is a structure in which dynamic information is layered on top of a static road map in accordance with the update frequency of each data. Figure 1 shows the structure of the dynamic map. The data obtained from the in-vehicle sensors are sent to the server that manages the dynamic map, and the application that achieves cooperative automatic driving runs on the basis of the data [13]. Therefore, dynamic information transmitted from the vehicle must always be sent to the server, and the server must process the information and send it with low latency to the vehicle [14]. In addition, the vehicle generally transmits to the dynamic map at 100-millisecond intervals [15]. However, the number of vehicles handled by the dynamic map is huge, and congestion is a concern if all vehicles continue to transmit data at high frequencies.

Therefore, congestion can be alleviated by adjusting the data transmission interval, considering the traffic environment around the vehicle. Although there is controversy over the arrangement of servers to manage dynamic maps [16]–[19], in this study, we constructed a platform for dynamic maps consisting of vehicles and servers. In addition, we have implemented a priority processing function that divides the lane where a vehicle travels into sections (Lane ID) on the basis of the traffic environment around the vehicle, sets the priority for each Lane ID, adjusts the data transmission interval from the vehicle in accordance with the priority, and evaluate its effectiveness.

In Section 2, we describe the basic structure of the dynamic map. In Section 3, we describe the priority processing function by adjusting the transmission interval. In Section 4, we explain how to determine the transmission interval considering the traffic environment. In Section 6 The effectiveness of the system is evaluated in Section 6. In Section 7, the results of the evaluation are discussed, and Section 8 is a conclusion.

II. Communication Method for Dynamic Map

In the dynamic map developed in this study, the data that a vehicle sends to the server includes vehicle ID, vehicle position, speed, time stamp, etc [20], [21]. The vehicle position is obtained using the vehicle ' s GPS position and scan matching [22]. A vehicle and a server are called nodes, which

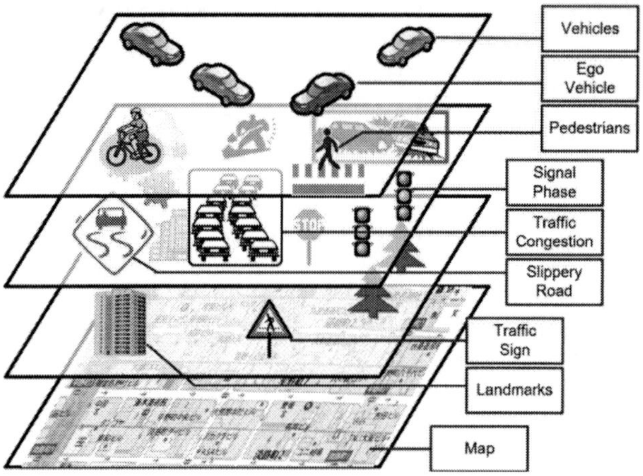

Figure 1. Overview of dynamic map

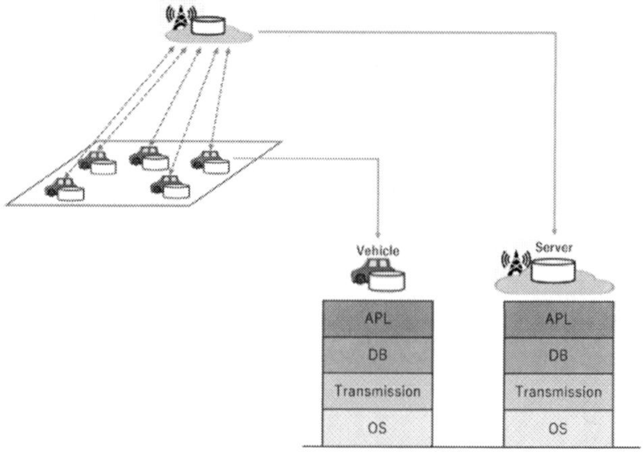

Figure 2. Dynamic map of server and vehicles

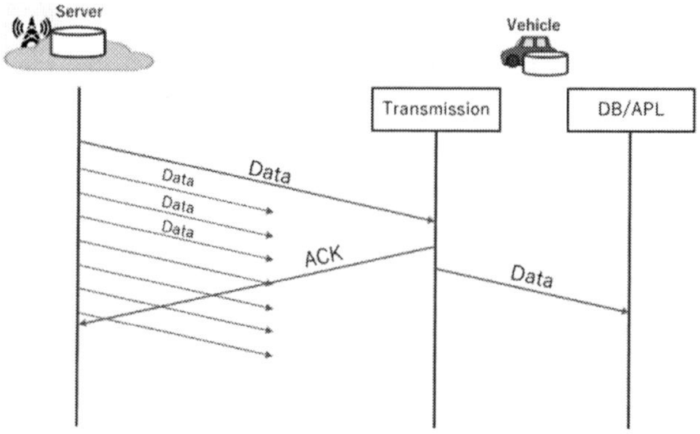

Figure 3. Sequence of resend function

are composed of four layers: an Operating System (OS), a Communication Unit (Transmission), a Database System (DB), and an Application (APL). Each node communicates through a communication section, and the data sent and received is processed by the application of each node. Figure 2 shows an overview of the constructed dynamic map.

The vehicle and the server need to send and receive data at a high speed, and in this study, UDP is used for communication [23]. However, although UDP has a small header size and can send and receive a lot of application data, there is no guarantee that the packets will reach the user because it is a connectionless protocol [24]. Therefore, it is necessary to guarantee the communication by performing retransmission processing on the application side, or the application must be able to tolerate packet loss. In addition, since the server and the vehicle use wireless communications, which is considered to be less reliable than wired communication, a function to send Acknowledgement (ACK) data was constructed.

As shown in Figure 3, the server sends data to the vehicle. The vehicle will then send an ACK to the server to confirm the received data. During this time, the server will continue to retransmit the data periodically until the ACK has been received. Once received, it will stop the retransmission. This enables the reliability of the communication to be maintained even with UDP.

III. Priority Processing by Adjisting the Transmission Interval

Safe-driving support applications, such as merging and mediation, require the position and speed information of each vehicle [25], [26]. Such applications require the high-frequency acquisition of location information for vehicles approaching or being within an intersection. However, location information for vehicles far from the intersection or moving away from it is not needed as frequently. However, every vehicle sends data to the server at regular intervals, regardless of the application's request. As a result, the processing and communication bandwidth of the server is tight, which may interfere with the services to support safe driving.

Therefore, we developed a function to minimize the impact on traffic and alleviate the processing load and bandwidth congestion on the server by setting the priority in accordance with the position of the vehicle on the road and adjusting the transmission interval from the vehicle by the communication section of the vehicle.

In Figure 4, the server receives data at regular intervals from vehicles over all the areas in the figure. Under this condition, it is not possible to filter out the vehicles because it is unclear which vehicles are placed in which traffic environment. Therefore, as shown in Figure 5, when vehicle location information is linked to road map information, it is possible to understand the traffic situation such as vehicles heading towards or away from an intersection. As a result, it is possible to prioritize each vehicle in consideration of the traffic environment. When the server receives data from the vehicle, it grasps the location information of the vehicle and notifies the vehicle of the transmission interval of the data in accordance with the priority in the response data. The vehicle transmits subsequent data at the transmission interval indicated by the server. By adjusting the transmission interval in accordance with the position of the vehicle, a priority processing function based on the data transmission interval from the vehicle in consideration of the traffic environment is achieved.

Figure 6 shows the sequence of the priority processing function by adjusting the data transmission interval of a vehicle. The vehicle sends data to the server through the communication section. The server creates ACK data from the received data and transmits it to the vehicle with the delay time for each Lane ID to control the transmission interval. The

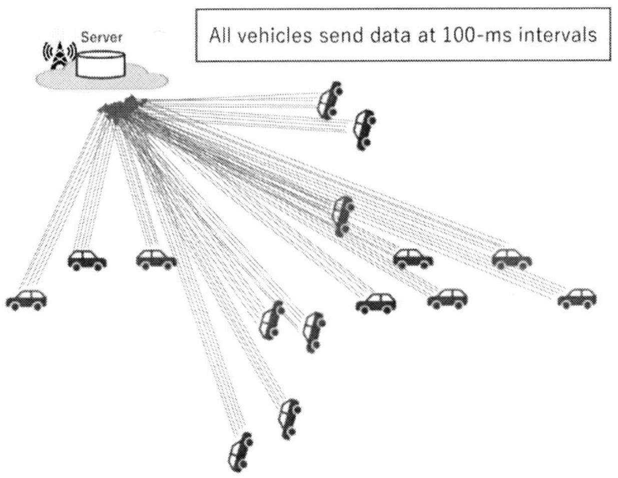

Figure 4. Communication traffic when transmission interval is constant

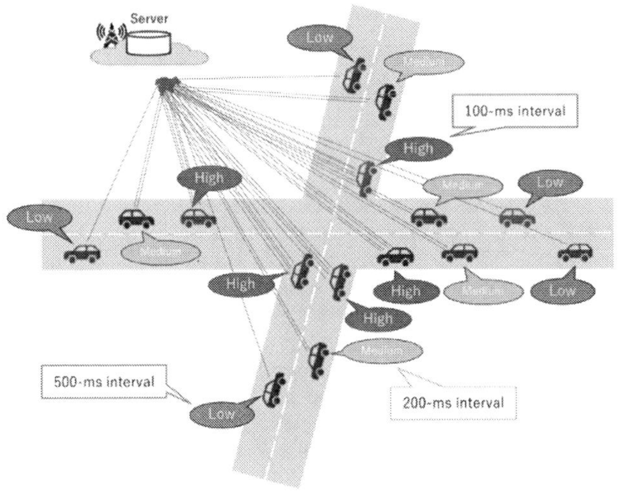

Figure 5. Adjustment of transmission interval in accordance with priority

vehicle adjusts its transmission interval in accordance with the delay time.

Figure 7 shows a flowchart of the priority processing function to adjust the data transmission interval of the vehicle. If the destination of the ACK data sent from the server is the vehicle, the transmission interval of the vehicle is adjusted in accordance with that in the ACK data.

IV. Prioritization in Consideration of the Traffic Environment

We determined the priority of the transmission interval on the road where a vehicle is traveling on the basis of the traffic environment around it. For example, suppose there was a road like the one shown in Figure 8. Since applications on dynamic maps process data from vehicles in real time, they need to transmit data at a high frequency in and around intersections. However, it is not necessary to transmit data at such a high frequency on roads that are far from intersections. Therefore, the lane is divided into sections (Lane ID) in accordance with the characteristics of the road on which the vehicle is traveling, and the transmission interval is determined for each Lane ID.

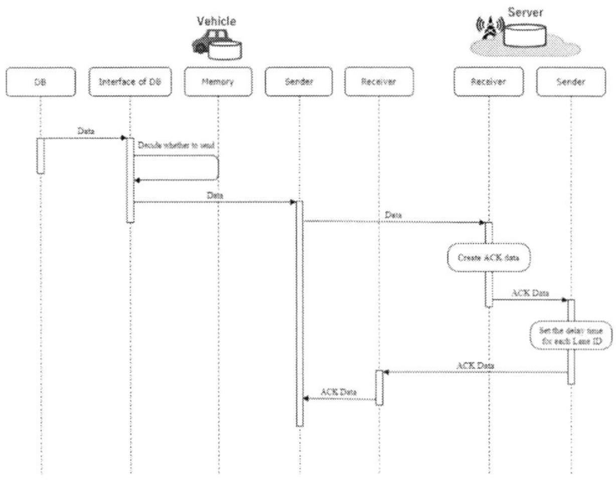

Figure 6. Sequence diagram of priority processing by transmission interval adjustment

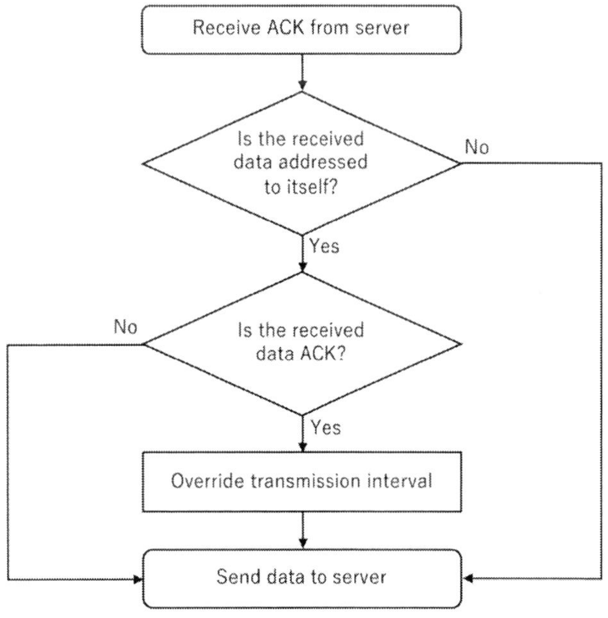

Figure 7. Flowchart of priority processing by transmission interval adjustment

A Lane ID is assigned to one lane of the road in Figure 8 for each road characteristic. The relationship between road characteristics and priority is shown in Table I. For example, if we want to apply a new road in the parking lot to the dynamic map, we can add the data to this database and set it as a Lane ID. The relationship between the priority and the transmission interval is shown in Table II. On the basis of these, the Lane ID is related to the transmission intervals. It is expected that the network will evolve and be able to transmit large amounts of data at higher speeds in the future. Therefore, if we want to communicate data more frequently, we can modify this database to briefly improve the communication interval of the entire dynamic map. The relationship between the Lane ID and the transmission interval is shown in Table III. The server determines the transmission interval to the vehicle based on

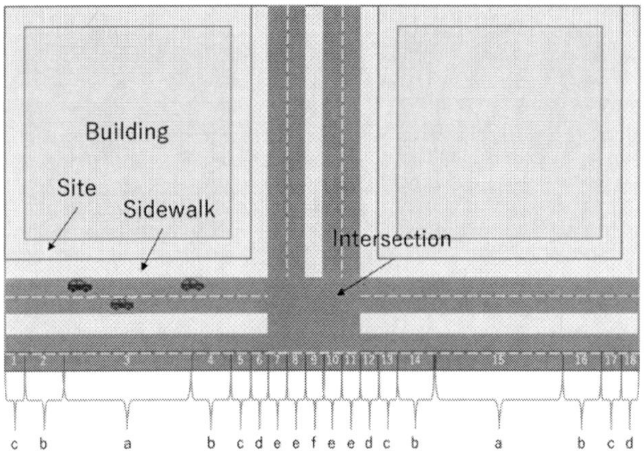

Figure 8. Priority corresponding to Lane ID

TABLE I. PRIORITY FOR ROAD CHARACTERISTICS

Symbol in the figure	*Road characteristics*	*Priority*
a	Building site center	1
b	Building site	2
c	Building site edge	3
d	Sidewalk / Side road	4
e	Lane	5
f	Intersection center	6

TABLE II. TRANSMISSION INTERVAL FOR PRIORITY

Priority	*Transmission interval*
1	500 ms
2	300 ms
3	200 ms
4	100 ms

TABLE III. TRANSMISSION INTERVAL FOR LANE ID

Lane ID	*Transmission interval*
1	200 ms
2	300 ms
3	500 ms
4	300 ms
5	200 ms
6	200 ms
...	...

this database. It is expected that the transmission intervals will be adjusted daily while operating a dynamic map, and it will be possible to operate the database concisely by normalizing the database and managing it independently. As the dynamic map utilizes this relationship, when the server receives data from a vehicle, it determines the appropriate transmission interval on the basis of the location information in the data, and transmits this information in the ACK to the vehicle. That way, the vehicle can adjust its transmission interval appropriately.

V. EVALUATION SYSTEM

To evaluate the priority processing function built in this study, a dynamic map system was constructed using two PCs to act as a server and vehicle, respectively. We have developed a dynamic map platform that covers everything from communication to applications, and we used that application to perform simulations in this study as well. However, the effectiveness of the proposed system is not clear due to the large amount of uncertainty in using this application. Also, a discussion of the internal behavior of the application is not the essence of this paper. Therefore, in this simulation, the data sent from the vehicle was assumed to be processed by the same application on the server and return an ACK. The vehicle acquires sensor information by the application and sends it to the server through the communication section. The data is temporarily stored in a queue on the server and processed by the application in turn. Table IV shows the specifications of the PCs used for the server and vehicle. Note that the synchronous method only applies to the vehicle PC. Also, since cars are considered to communicate with the dynamic map wirelessly, we have conducted a demonstration experiment using wireless communication by placing an edge server at a mobile phone base station. However, the proposed system is a dynamic map system, and the use of wireless communication is highly dependent on the communication method and conditions, and the uncertainties are large. Our dynamic map platform can be applied to any communication method, and better communication methods can be adopted as the network evolves. Therefore, in order to evaluate the proposed system in detail, the vehicle and the server were connected by a wired connection, eliminating the uncertainties of wireless communication.

TABLE IV. SERVER AND VEHICLE CONFIGURATION

OS	Ubuntu 16.04
CPU	8-core 16-thread (3.60 GHz)
Memory	16 GB
SSD	256GB
Communication method	Wired (Up to 1 Gbps)
Synchronous method(Vehicle only)	No

TABLE V. SIMULATION CONDITIONS

Number of lanes	56
Maximum number of vehicles per lanes	89
Speed	40 km/h
Vehicle length	4.7 m
Distance between vehicles	20 m
Total number of vehicles	4984

Figure 9 shows the road map used for the evaluation. We used the Manhattan model, which consists of alternating two-lane streets on one side and four-lane streets on the other. There are 56 lanes within a square range of 2205 m per side, with vehicles running at regular intervals in the opposite direction from the end of each lane. The specifications for this simulation are shown in Table V. The length of the vehicle was set to 4.7 m, and the distance between vehicles was set to 20 m.

VI. RESULT

A. *Number of vehicle data to be sent and received*

Figure 10 shows a comparison of the peak number of the data received by the server. We compared three possible instances. First, there is no retransmission by ACK and the transmission interval from the vehicle remains fixed at 100 ms. Second, there is a retransmission but the transmission interval remains fixed at 100 ms. Third, there is a retransmission and the transmission interval is adjusted by priority processing. By using the priority processing function, we were able to suppress the peak rate of the data received by the server to

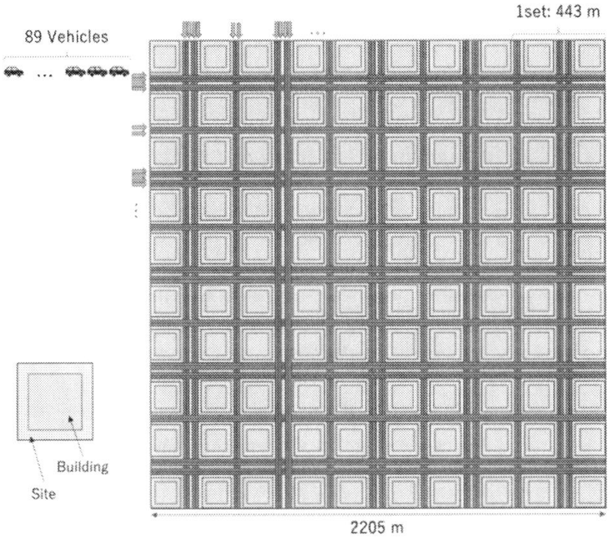

Figure 9. Road map used in the simulation

TABLE VI. PACKET LOSS RATIO

System	*Packet loss ratio*
No Resend & No Priority	0.006 %
Resend & No Priority	0 %
Resend & Priority	0 %

about 70 vehicles. Also, the number of data received by the system with resend is higher than the system without resend. In the case of no resend, this is because the vehicle is sending data at a high frequency and part of it is causing packet loss. Therefore, a resend function is used, and the number of data received is slightly increased. The packet loss rate for each system is shown in Table VI.

We also evaluated the rate at which the server sends and receives data. Figure 11 shows the reception and transmission rates when there is a retransmission with and without priority processing, respectively. The horizontal axis shows the elapsed time from when the first vehicle entered the road in the evaluation range. The vertical axis shows the rate of how much data the server is receiving and sending per second. In the absence of priority processing, the server cannot keep up with the data received from the vehicle, resulting in a processing delay. Therefore, the retransmission is not finished even after 400 seconds of transmission from the vehicle, and the convergence takes a long time. However, by using priority processing, we were able to reduce the rate of data transmission and reception in the server, in which both ended as soon as the transmission from the vehicle was completed without any processing delay.

B. *Processing Latency and Scalability*

The scalability of the three systems was evaluated by varying the number of lanes and the number of vehicles. In the Manhattan model presented in Section V, the total number of vehicles running during the simulation is 4984. In this case, no processing delay occurred in the system with priority processing, but processing delay occurred in the system without priority processing. Therefore, by varying the

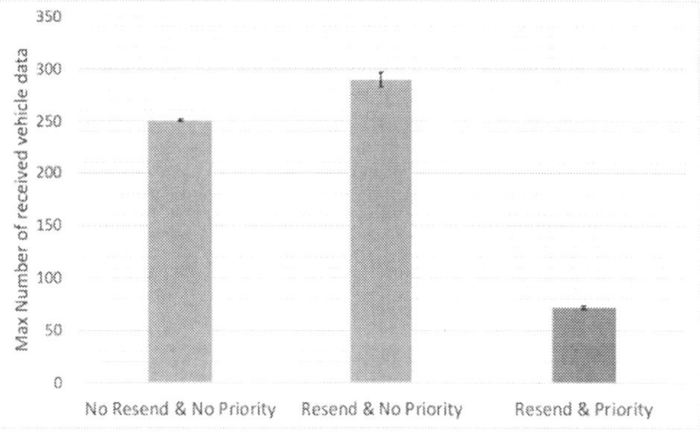

Figure 10. Maximum rate received by server

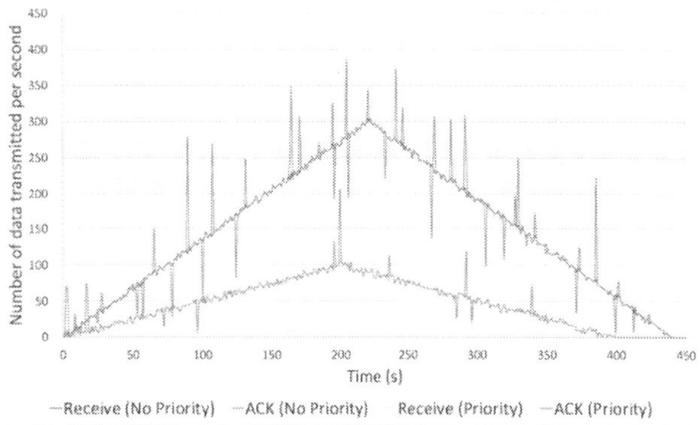

Figure 11. Processing delay time

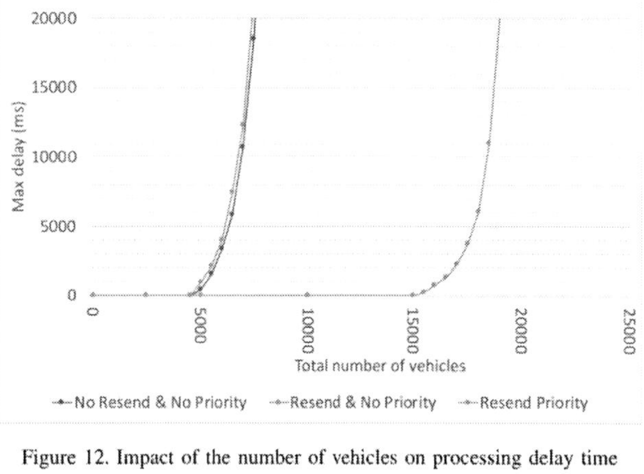

Figure 12. Impact of the number of vehicles on processing delay time

number of vehicles and the number of lanes, we adjusted the total number of vehicles in the simulation and evaluated the maximum processing delay for each. The evaluation results are shown in Figure 12. Scalability was greatly improved by using the priority processing function, which enabled us to process about 15,000 vehicles with low latency. The reason why the scalability is almost the same with respect to systems without the priority processing function, regardless of the presence or absence of resend, is that the packet loss rate remains very low even in systems without resend.

VII. Discussion

As shown in Figure 10, the amount of data transmitted by the vehicle is higher with retransmissions than without retransmissions. However, even if there is a retransmission process, the data volume can be significantly reduced by using priority processing, because the data can be reduced by adjusting the transmission interval on the basis of the Lane ID.

In addition, as shown in Figure 11, due to the processing delay and lack of priority processing, the server continues to receive data from the vehicle after about 400 s when transmissions other than retransmissions have completed. This is because the transmission interval of data from the vehicle is fixed at 100 ms, and the processing delay is caused by receiving data that exceeds the processing performance of the server. However, in the case of priority processing, the transmission and reception of the server ended at the same timing as the data transmission from the vehicle was completed, and no processing delay occurred. In Figure 12, the amount of data that the server needs to process is greatly reduced by using the priority processing function, which greatly improves the scalability of the dynamic map.

Dynamic maps for safe-driving support and automatic driving need to be communicated and processed with low latency. In addition, the server must be able to reliably receive data from a vehicle. Furthermore, the number of vehicles communicating with the server is expected to increase in the future. By using the retransmission function to ensure the reliability of the communication between a vehicle and server, and by using the priority processing function by adjusting the transmission interval, the amount of data received by the server can be reduced to about one-fifth of that of a system with a fixed transmission interval. This will also reduce processing delays and lead to smoother traffic flow.

VIII. Conclusion

In recent years, research and development for automatic driving has attracted much attention, but the range of recognition is limited due to the limitations of in-vehicle sensors. Therefore, research has begun on cooperative automatic driving, in which automatic vehicles share data obtained from sensors, etc., with the aim of improving safety and efficiency. In addition, dynamic maps, a common information and communication platform for the integrated management of shared sensor information, are under consideration. A vehicle always sends data to a server that manages the dynamic map, and the server runs applications for driving support and control on the basis of the data, so fast information processing is required. However, if data is continuously sent from vehicles to the server at high transmission intervals and many vehicles are managed by the dynamic maps on the server, communication congestion and processing load becomes a concern. In addition, the transmission interval of data from a vehicle required by the road characteristics varies in actual traffic environments. Therefore, congestion can be alleviated by adjusting the transmission interval of data from the vehicle in consideration of road characteristics. In this paper, a platform for a dynamic map consisting of a server and a vehicle is constructed. By implementing the retransmission function, we have achieved highly reliable communication even for UDP. In addition, a priority processing function that adjusts the transmission interval is implemented by setting the priority for each section of the lane (lane ID) where a vehicle is traveling on the basis of the road characteristics around the vehicle. We evaluated the amount of data sent and received by the server when there were no retransmissions, when there were resend and no priority processing by adjusting the transmission interval, and when there were resend and priority processing by adjusting the transmission interval. As a result, the maximum reception rate could be reduced by about 80%. We also measured the processing delay of the server and showed that it can be processed with low latency. These results show that the function built in this paper is effective in improving the efficiency of communication between vehicles and servers in dynamic maps.

References

[1] A. Geiger, P. Lenz and R. Urtasun, "Are we ready for autonomous driving? The KITTI vision benchmark suite," 2012 IEEE Conference on Computer Vision and Pattern Recognition, 2012, pp. 3354–3361.

[2] J. Baber, J. Kolodko, T. Noel, M. Parent and L. Vlacic, "Cooperative autonomous driving: intelligent vehicles sharing city roads," IEEE Robotics & Automation Magazine, vol. 12, no. 1, 2005, pp. 44–49.

[3] L. Hobert, A. Festag, I. Llatser, L. Altomare, F. Visintainer and A. Kovacs, "Enhancements of V2X communication in support of cooperative autonomous driving," IEEE Communications Magazine, vol. 53, no. 12, 2015, pp. 64–70.

[4] J. Zhang, F. Wang, K. Wang, W. Lin, X. Xu and C. Chen, "Data-Driven Intelligent Transportation Systems: A Survey," IEEE Transactions on Intelligent Transportation Systems, vol. 12, no. 4, 2011, pp. 1624–1639.

[5] G. Dimitrakopoulos and P. Demestichas, "Intelligent Transportation Systems," IEEE Vehicular Technology Magazine, vol. 5, no. 1, 2010, pp. 77–84.

[6] J. Lee and B. Park, "Development and Evaluation of a Cooperative Vehicle Intersection Control Algorithm Under the Connected Vehicles Environment," IEEE Transactions on Intelligent Transportation Systems, vol. 13, no. 1, 2012, pp. 81–90.

[7] S. Bowles and H. Gintis, Ed., A Cooperative Species: Human Reciprocity and Its Evolution. Princeton University Press, 2011.

[8] ETSI, "Intelligent Transport Systems (ITS); V2X Communications; Multimedia Content Dissemination (MCD) Basic Service specification; Release 2," 2019, TS 103 152 V2.1.1.

[9] ——, "Intelligent Transport System (ITS); Users and applications requirements; Part 2: Applications and facilities layer common data dictionary," 2014, TS 102 894-2 v1.2.1.

[10] J. Leonard, H. Durrant-Whyte and I. J. Cox, "Dynamic map building for autonomous mobile robot," IEEE International Workshop on Intelligent Robots and Systems, Towards a New Frontier of Applications, 1990, pp. 89–96.

[11] H. Shimada, A. Yamaguchi, H. Takada and K. Sato, "Implementation and Evaluation of Local Dynamic Map in Safety Driving Systems," Journal of Transportation Technologies, vol. 5, no. 2, 2015, pp. 103–112.

[12] "Dynamic Map 2.0 Consortium." [Online]. Available: {http://www.nces.i.nagoya-u.ac.jp/dm2/}[accessed:2020-08-12]

[13] K. Sato, Y. Watanabe and H. Takada, "Dynamic Map as Common Application Platform for Dynamic Geographic Information Management," The journal of the Institute of Electronics, Information and Communication Engineers, vol. 101, no. 1, 2018, pp. 85–90.

[14] S. Steven and K. Thomas, "Traffic probe data processing for full-scale deployment of vehicle-infrastructure integration," Transportation research record, vol. 2086, no. 1, 2008, pp. 115–123.

[15] NTT Docomo and Pasco, "Realization of efficient updating and distribution of advanced map database." [Online]. Available: {https://smartiot-forum.jp/application/files/6414/7702/6769/sympo_20160927_02_mobility_03-03.pdf}[accessed:2020-08-12]

[16] ITS Information and Communication System Promotion Conference Cellular System TG, "Toward advanced ITS and autonomous driving using cellular communication technology Survey Report."

[Online]. Available: {https://itsforum.gr.jp/Public/J7Database/p62/Cellular_system_201906.pdf}[accessed:2020-08-12]

[17] ETSI, "Multi-access Edge Computing (MEC); Study on MEC Support for V2X Use Cases," 2018, GR MEC 022 v2.1.1.

[18] ——, "Intelligent Transport System (ITS); Vehicular Communications; Basic Set of Applications; Part 2: Specification of Cooperative Awareness Basic Service," 2018, EN 302 637-2 v1.4.0.

[19] ——, "Intelligent Transport System (ITS); Vehicular Communications; Basic Set of Applications; Part 3: Specifications of Decentralized Environmental Notification Basic Service," 2014, EN 302 637-3 v1.2.1.

[20] C. Nanthawichit, T. Nakatsuji and H. Suzuki, "Application of Probe-Vehicle Data for Real-Time Traffic-State Estimation and Short-Term Travel-Time Prediction on a Freeway," Journal of the Transportation Research Board, vol. 1855, no. 1, 2003, pp. 49–59.

[21] S. E Shladover and T. M Kuhn, "Traffic Probe Data Processing for Full-Scale Deployment of Vehicle-Infrastructure Integration," Journal of the Transportation Research Board, vol. 2086, no. 1, 2008, pp. 115–123.

[22] K. Takagi, K. Morikawa and T. Ogawa, "Road Environment Recognition Using On-vehicle LIDAR," 2006 IEEE Intelligent Vehicles Symposium, 2006, pp. 120–125.

[23] J. Postel, "User Datagram Protocol," 1980, RFC 768.

[24] C. Shue, W. Haggerty and K. Dobbins, "OSI Connectionless Transport Services on top of UDP Version: 1," 1991, RFC 1240.

[25] U. Franke, D. Gavrila, S. Gorzig, F. Lindner, F. Puetzold and C. Wohler, "Autonomous driving goes downtown," IEEE Intelligent Systems and their Applications, vol. 13, no. 6, 1998, pp. 40–48.

[26] M. Gerla, E. Lee, G. Pau and U. Lee, "Internet of vehicles: From intelligent grid to autonomous cars and vehicular clouds," 2014 IEEE World Forum on Internet of Things (WF-IoT), 2014.

Secure Routine

A Routine-Based Algorithm for Drivers Identification

Davide Micale*, Gianpiero Costantino†, Ilaria Matteucci†, Giuseppe Patanè‡ and Giampaolo Bella*
*University of Catania, Dip. di Matematica e Informatica, Catania, Italy
†Consiglio Nazionale delle Ricerche (CNR), Istituto di Informatica e Telematica (IIT), Pisa, Italy
‡Park Smart Srl, Catania, Italy
Email: davide.micale@phd.unict.it, name.surname@iit.cnr.it, giuseppe.patane@parksmart.it, giamp@dmi.unict.it

***Abstract*—The introduction of Information and Communication Technology (ICT) in transportation systems leads to several advantages (efficiency of transport, mobility, traffic management). However, it may bring some drawbacks in terms of increasing security challenges, also related to human behaviour. As an example, in the last decades attempts to characterize drivers' behaviour have been mostly targeted. This paper presents *Secure Routine*, a paradigm that uses driver's habits to driver identification and, in particular, to distinguish the vehicle's owner from other drivers. We evaluate Secure Routine in combination with other three existing research works based on machine learning techniques. Results are measured using well-known metrics and show that Secure Routine outperforms the compared works.**

Keywords–driver identification; secure routine; machine learning; automotive.

I. INTRODUCTION

Modern vehicles can be considered as computer on wheels. The mechanical parts are often controlled by software components and communication protocols are in charge of exchanging data among vehicle's components. For this reason, modern vehicles are Cyber Physical Systems *(CPS)* in which used technologies bring countless advantages in terms of, for instance, efficiency of city operations and services. An example among all is the Internet connectivity. Within this context, a problem of particular interest is how to leverage vehicular and/or smartphone data to characterize driver identification. Its characterization finds application in the development of software, which can be used by insurance companies to check and identify drivers or, for instance, to discourage auto theft. In 2019, around 56k vehicles were targeted by thieves in UK [1]. It equates to one car stolen every 9 minutes and 45% of thefts occurred between midnight and 6 AM. Having a strategy to classify the driver's behaviour may help to mitigate this trend.

Routine based classification is a type of classification [2] that aims to find actions that are frequently repeated in time. To complete a task, people repeat sequence of actions previously saw from others or done by themselves, no matter how tough the task is [3]. Two persons may accomplish the same task with similar actions but with little fundamental differences [3]. Routines can describe how people organize their lives: daily commute, weekly, meetings, holidays. Routines can also describe how a driver approaches to an intersection [4].

Based on these aspects of routine, here we introduce the paradigm of Secure Routine *(SR)* that takes into account not only what the user does but also how much frequently. We use the SR paradigm within the automotive context with the aim to classify drivers. To achieve this, we elaborate and implement the SR algorithm that exploits sensors' car data, obtained, for instance, through the *OBD-II* [5] diagnostic port. The SR algorithm evaluates the recorded data and, in particular, uses the timestamp to make an accurate classification of drivers. Then, SR leverage a Machine Learning *(ML)* technique to establish driver's routines and to properly identify the driver.

To test the goodness of the Secure Routine algorithm, we compare it with other research works present in literature. The comparison is done on two different datasets and the results are evaluated using three metrics: *Accuracy*, *Precision* and *Recall*. Findings show that Secure Routine outperforms the compared works in all the tests carried out.

The paper is structured as follows: next section presents the state of the art. In Section III, we introduce the background on ML techniques. In Section IV, we present the Secure Routine paradigm used to identify drivers. Then, in Section V, we compare Secure Routine with other research works presented in literature. Finally, Section VI draws the conclusion of this paper and presents some hits for future research directions.

II. STATE OF THE ART

In literature, there are several solutions based on ML techniques for the identification of driver's behaviour. Bernardi et al. [6] used a Multi Layer Perception (MLP) to identify drivers. They used three datasets obtaining respectively 94%, 95% and 92% of Accuracy. In particular, these results were obtained using a Start&Stop sliding window. A sliding window combines several consecutive instances in a single instance. In particular, Start&Stop joins instances starting when the car is moving until the car stops.

Gao et al. [7] discriminated drivers through Stop-and-Go events using a *voting strategy*. A Stop-and-Go event occurs when the car slowdowns until stops (stop phase), it stands still for five or more seconds and then speeds up (go phase).

Wang et al. [8] identified 30 drivers by using the voting strategy and Random Forest algorithm. Authors split data and tests into different window sizes. They use six sensor signals and three derived sensor's signals along with five statistical features. With 5 minutes of testing data this model achieves almost 93% of Accuracy. With a sliding window of 5 seconds and 6 minutes of testing data they achieve 100% of Accuracy.

Girma et al. in [9] used the Long Short-Term Memory (LSTM) algorithm with sliding windows and tested their model on [10] and [11] datasets with Precision and Recall of 98%.

Kwak et al. in [12] selected 15 features to identify drivers behaviour. For each feature they computed the mean, median and standard deviation according to a reference sliding window. Thus, the total number of features is 45. They used different ML algorithms and achieved the best Accuracy of 99,6% applying Random Forest on [10] dataset.

Martinelli et al. in [13] tested several Decision Tree algorithms with the same dataset [10] using all 51 features. They obtained a Precision and Recall equal to 99,2% with J48. The same authors in [14] used only six features out of 51 features of [10] dataset. In this case, Precision and Recall decreased to 98,9% due to under-fitting.

Compared to our paper, [6], [7], [8], [12], [13] and [14] do not look for frequency. Also, LSTM in [9] obtained lower scores in comparison with a *Decision Tree (DT)* algorithm ([12], [13] and [14]) on the same dataset. As shown by [14], certain features discriminate better than others for some drivers. Hence, SR must use the best feature set for each driver. [6], [14] and [13] are the only ones that make owner-driver identification, they select the same feature set for all drivers. Finally, SR breaks down the timestamp in fine grained units to detect frequency in order to increase the accuracy.

III. MACHINE LEARNING

ML is the study of computer algorithms that improve automatically through experience. ML algorithms build a mathematical model to make predictions or decisions without being explicitly programmed to do so. At the basis of the model, there is a dataset that has to be processed. Such dataset can be considered as a table in which all data are listed. Each row of the table is called *instance* and each column represents a *feature* of the instance. The dataset is usually split into two parts, the *training dataset* and the *test dataset*. The model is created on the basis of the training dataset. Instead, a *test dataset* represents all instances adopted to verify how much accurate our model is in doing the classification.

ML techniques are largely adopted for the identification and classification of users. In the following, we introduce an example of ML algorithm based on DT predictive modelling approach. A DT consists on a tree data structure that contains rules to classify the instance. For each level of the tree, the value of a feature of the instance is tested, for example, through a specific question. Each internal node of the tree contains a test. Depending on the answer, the model follows a different edge: the left edge if the result of the test is true, otherwise the right edge is followed. Finally, the leaf nodes, i.e., the nodes with no children, contain the prediction.

A. Decision Tree Requirements

A DT algorithm must create a tree with the minimum number of levels. This allows the ML algorithm to classify the instance as fast as possible. To build a DT with a low number of levels, it is necessary to select the best tests for the model. This is done by selecting the appropriate *Formula* to make the selection. A *Formula* specifies the criterion chosen to establish which is the next test to perform in the DT.

For instance, let *Alice* and *Bob* be two drivers that are used to going on the Sixth Avenue. Alice goes on the Sixth Avenue all days of the week, instead Bob goes only from Monday to Friday. Bob drives slightly faster than Alice, with a speed up to 55 Km/h. A possible DT model is the one in Figure 1(a) that is built by putting on the tree root the following test:

"Is today Saturday or Sunday?"

Following the root test, we have that the left child is taken by Alice instead the right child corresponds to the following test:

"Is the vehicle speed lower than 55 Km/h?"

Again the left child is a leaf node that represents Alice, whereas the right child is the leaf node representing Bob.

Despite the above DT model is a valid model for our example, we may produce a better tree in which a root node is configured with the following test (Figure 1(b)):

"Is the vehicle speed lower than 55 Km/h?"

In this case, the left child is the leaf node Alice and the right child is the leaf node Bob. Hence, a ML algorithm concludes its prediction with only one test.

A DT has to be simple. This allows the DT to be flexible enough to represent also further instances. Thus, if the built model is too complex, it may not represent new labelled instances, i.e., for instance those ones present in a test-set. This may cause a high error rates, generating the *over-fitting* error. To reduce the over-fitting error, the *pruning* technique can be adopted to obtain a simpler version of the tree by pruning some nodes. Another solution to mitigate the over-fitting error is the *feature selection* that works by removing features. However, pruning too many nodes and removing too many features or relevant ones may lead to higher error rates, aka *under-fitting*.

B. Decision Tree Algorithms

Several DT algorithms were developed to generate models. The *C4.5* was proposed in 1993 [15] and it uses the Gain Ratio (GR) of a feature "X" of the training set (T) to establish which is the best test to perform.

$$GR = \frac{H(T) - H(T|X)}{H(X)} \tag{1}$$

where:

- $H(T)$ indicates the *entropy* of T, i.e., the quantity of information carried by the probability distribution of labels in T [16], calculated as:

$$H(T) = -\sum_{j=1}^{k} \frac{freq(C_j,T)}{|T|} \times log_2\left(\frac{freq(C_j,T)}{|T|}\right) \tag{2}$$

  where:
  - ◦ k is the number of classes;
  - ◦ $freq(C_j,T)$ is the number of instances in the j-th class;
  - ◦ $|T|$ is the number of instances of T.
- $H(T|X)$ indicates the entropy after partitioning T in "n" parts, where "n" is the number of possible values assumed by X:

$$H(T|X) = \sum_{i=1}^{n} \frac{|T_i|}{|T|} \times H(T_i) \tag{3}$$

  where:
  - ◦ $|T_i|$ is the number of instances with the i-th value assumed by the feature X;
  - ◦ $H(T_i)$ indicates the entropy of the set of instances with the i-th value assumed by the feature X.
- $H(X)$ indicates the entropy of X:

$$H(X) = -\sum_{i=1}^{n} \frac{|T_i|}{|T|} \times log_2\left(\frac{|T_i|}{|T|}\right) \tag{4}$$

(a) DT with two levels.

(b) DT with one level.

Figure 1. Comparison of two possible DT for solving the same problem.

Note that C4.5 can handle features with unknown values and real numbers and may make use of the pruning technique.

Random Forest (RF) [17][18] is an algorithm formed by a set of DTs. Each tree is built from a random sampling with replacement of the training-set. Each node of a tree is the best test defined on a subset of features, instead of on all available ones. Trees are not pruned. In prediction phase, an instance is run on each tree and each tree makes a prediction. The most predicted value becomes the prediction of RF. Also, RF includes a procedure in case of unknown values in the dataset.

IV. Secure Routine

In literature, the concept of *Routine* is already exploited to classify users or drivers [4]. A Routine is defined as a set of actions that a person frequently perform in response to a circumstance [19]. Hence, routines can describe how people organize their lives: daily commute, weekly, meetings, holidays. Here, we refine the concept of Routine by introducing the paradigm of *Secure Routine* that takes into account not only what the user does but also how much frequently.

We define SR and present its application into the automotive context to perform driver's behavioural identification. To this aim, SR analyses all tracking data recorded by vehicle's sensors while the user is driving it. Tracked data are organized in separate instances according to the sensor that collects them and the timestamps when the event occurs. Hence, SR firstly decomposes the timestamp of each instance and extracts second, minute, hour, day of week, day, month and year. Then, SR removes less relevant features, as we will describe below using the Feature Selection *(FS)* technique. Successively, the data collected by sensors are correlated with the timestamp previously decomposed. Then, a ML algorithm examines these data. The output is a model representing users' Secure Routine. As final step, the obtained model is compared with an observed user's driver behaviour for his/her identification.

To show the value added by the Secure Routine to identify drivers, we introduce the following example. Let us consider Alice and Bob who are used to going on the Sixth Avenue. Alice usually goes there at 12PM, and Bob at 7PM. If we do not consider the timestamp information, the resulting model of Alice and Bob will contain only the information *"The user is used to going on the Sixth Avenue"*. In this situation, the observed behaviour will be compared to understand whether the driver is Alice or Bob. However, this selection is quite difficult since the missing timestamp information is fundamental to distinguish between the drivers. On the contrary, if we consider also the timestamp in which the event happens, the identification will be unique in this case. In fact, if the vehicle is at 7PM on the Sixth Avenue, therefore the driver is Bob.

This is what Secure Routine does considering daily routines as well as monthly and yearly ones. Hence, SR may be very useful, for instance, to mitigate scenarios as the one depicted in Section I: in UK cars are often stolen at night. If the vehicle's owner does not usually drive during the night, SR can easily detect the weird behaviour. In particular, the SR paradigm is built upon a ML algorithm that uses as training-set the data recorded through an OBD-II device. A closer working mechanism of SR is presented in [20]. Here, the authors prefer to involve the interval between a rerun of the same action. Let us consider this other example in which Alice goes on the Sixth Avenue every 24 hours for the whole week, instead Bob every 24 hours from Monday to Friday. In this case, the routine of Bob will be modelled as intervals of 24 and 72 hours. So, if we consider a driver moving on Saturday, we would not be able to identify the driver, neither Alice nor Bob, since the interval is set to 24 hours. On the contrary, if the day of the week is taken into account, Alice will be correctly identified.

A. SR Algorithm

Let us consider a target vehicle belonging to a driver d. The SR algorithm acts in four phases:

a) Model Generation Dataset: Whenever a vehicle is used, its sensors register pieces of information about several *features*, e.g., the water temperature, the speed, the brake pressure, and so on. We assume to take trace of all these data in combination with the timestamp in which each instance of data is generated. Data are taken from the OBD-II port by using an OBD-II interface [21]. Each instance of data is called *interaction* of the driver d with the vehicle and it is denoted as $in_{i,d}$ where i is the timestamp. Interactions are composed by the timestamp, recorded with the following template: (day, month, year, hour, minute, second and day of the week) plus the others features obtained from the OBD-II.

b) FS paradigm: To mitigate the possible over-fitting error, we implement the *FSParadigm* (Figure 2).

FSParadigm is designed to select the best features to use. It firstly ranks all features applying the *Gain Ratio* approach and then features are sorted in ascending order. Those features with rank equal to zero are discarded. Then, the average-rank among all features not correlated to the timestamp is calculated. The FS discards those features, except those related to time, whose rank sum is less than or equal to the average-rank.

c) Model Generation Algorithm: Let us consider that a vehicle may be driven by d but also by other people, e.g., friends or relatives of d. In the modelling phase, our algorithm

(Figure 3) considers all the past interactions recorded by the vehicle and labels with 1 each interaction that belongs to d, 0 otherwise. The labelled interactions are sent to a DT algorithm that generates the model for the driver d.

In particular, in line 5, *FSParadigm* is the Feature Selection paradigm we described above as part of SR and line 6 (*MLAlgorithm*) indicates the ML algorithm in use with the subset of features obtained before.

d) SR Identification strategy: Once the model is generated, SR makes the identification evaluating each interaction. In particular, SR links an interaction to the vehicles' owner if the ML algorithm predicts and labels it as 1, otherwise 0.

V. SECURE ROUTINE EVALUATION

We evaluated Secure Routine in two steps: first, we run it using two ML algorithms and we verified which of them best performs to identify drivers. Then, we compared Secure Routine with the following research works present in literature:

- Martinelli et al. [14] referred in the following as M.
- Kwak et al. [12] referred in the following as K.
- Girma et al. [9] referred in the following as G.

A. Datasets

We run the experiments using two datasets presented in [10], referred as Θ, and [22], referred as Ψ. The former is a dataset used also by M, K and G in their research works. So we can fairly make a comparison. However, the Θ dataset does not contain a fundamental feature used by SR, this is the *timestamp* of each represented instance. Nevertheless, Θ dataset contains the *engine runtime* that provides the minutes to be used as timestamp needed for SR to work.

On the other hand, Ψ dataset contains a timestamp for each instance by default. This feature allows Secure Routine to fully work by using all available pieces of information. In particular, SR expands the timestamp to generate all time dependent features. As far as we know, the other compared research works do not make use of this dataset to evaluate their proposal. So, to evaluate SR even in this case, we were able to re-run the work proposed by Martinelli et al. and calculate the results for the owner-driver identification. On the other side, the works K and G did not calculate the owner-driver identification and, also, it was not possible to re-run their algorithms since the implementation is not publicly available. In the specific case of G, the authors published only the pre-built model and we were not able to use it.

B. Metrics

To get a comparable result of SR with M, K and G, we evaluate *Accuracy* [23], *Precision* and *Recall* [14].

- *Accuracy* represents how often the model is making a correct prediction. It is the ratio between the number of correct predictions and the number of predictions:

$$Accuracy = \frac{TP + TN}{TP + TN + FP + FN} \tag{5}$$

  where:
  - TP (True Positive) is the number of instances belonging to the vehicle's owner that are correctly predicted;
  - TN (True Negative) is the number of instances not belonging to the vehicle's owner that are correctly predicted;
  - FP (False Positive) is the number of instances belonging to another person but incorrectly predicted;
  - FN (False Negative) is the number of instances belonging to the vehicle's owner but incorrectly predicted.
- *Precision* measures how often the predicted instances belonging to the vehicle's owner are true. It is calculated as the ratio between TP and $TP + FP$:

$$Precision = \frac{TP}{TP + FP} \tag{6}$$

- *Recall* identifies how often the instances belonging to vehicle's owner are correctly predicted. It is calculated as the ratio between TP and $TP + FN$:

$$Recall = \frac{TP}{TP + FN} \tag{7}$$

To better estimate the three metrics depicted above, in our experiments we used the 10-fold cross-validation [24] approach. First, we split the dataset on 10 equal size subsets D_1, D_2, ..., D_{10}. Each instance of the dataset is randomly inserted in a subset. Then, we constructed 10 training sets Tr_1, Tr_2, ..., Tr_{10} and 10 testing sets Te_1, ..., Te_{10}. Tr_i is made of all subsets except D_i and Te_i is made of D_i

```
function FSParadigm(instances)
    ranking ← GR(instances)
    ranking_ordered ← order ranking ascending
    features_>0 ← discard features with rank =
        0 from ranking_ordered
    (features_no_timestamp_correlated,
        features_timestamp_correlated) ← features_>0
    ranking_no_timestamp_correlated ← ranking from
        ranking_ordered of features present in
        features_no_timestamp_correlated
    average_ranking ←
        mean(ranking_no_timestamp_correlated)
    subset_no_timestamp_correlated ← discard
        features sum is less than or equal to
        the average_ranking from
        ranking_no_timestamp_correlated
    subset ← subset_no_timestamp_correlated ∪
        features_timestamp_correlated
    return subset
```

Figure 2. Feature Selection Paradigm

```
1   function generate_model(d)
2       ins_d ← get interactions from db made by d,
            labeling 1
3       ins_o ← get interactions from db made by
            others, labeling 0
4       ins_all ← ins_d ∪ ins_o
5       subset ← FSParadigm(ins_all)
6       model ← MLAlgorithm(ins_all with features
            from subset)
7       return model
```

Figure 3. Secure Routine Model Generation

with $i \in \{1, 2, ..., 10\}$. For each pair (Tr_i, Te_i) is calculated $Accuracy_i$, $Precision_i$ and $Recall_i$. Finally, we calculated the final value of $Accuracy$, $Precision$ and $Recall$ as the mean of $Accuracy_i$, $Precision_i$ and $Recall_i$, respectively.

C. Experiments

We performed four types of experiments to evaluate Secure Routine. The first experiment is related to multi-driver identification problem [14], i.e., properly identify who is the driver. However, as step zero, we decided to find the most suitable ML algorithm with the best features set to evaluate SR. We leverage on Weka [25] as software that contains a collection of visualization tools and algorithms for data analysis and predictive modelling. So, we used the available Gain Ratio method to rank each feature. Then, we employed *J48*, which is the implementation of the C4.5 algorithm, and RF algorithm over the driver identification.

In this step, results are obtained on Θ dataset. It contains data from 10 drivers. Figure 4(a) shows the driver instances' distribution. Drivers have 9438 instances on average: Driver 4 has the highest number of instances with 13244 samples while Driver 1 has the lowest number with 7240 instances. In addition, drivers drove two times in the same path in similar time-window. Dataset instances are recorded per second.

Table I shows the results obtained comparing SR implemented into J48 and RF algorithms applied to the driver identification using Θ dataset. RF algorithm with feature selection (37 features) obtained the best Precision and Recall.

After selecting SR with RF and the most appropriate features ranked by the Gain Ratio method, we show the first experiment results obtained by comparing SR with the work in M, K and G on the Θ dataset. As shown in Table II(a), Secure Routine and K achieves the best results. Note that M did not calculate the accuracy in the paper, so we established this value through the replication of their experiment. Instead, K did not provide on their research Precision and Recall. Finally, for G we were not able to retrieve the exact Accuracy.

As we can see in Table II(b), SR achieves almost a perfect Precision, i.e., 100%, but with the worst Recall and this depends on the features selection. In fact, if we increment the number of features, we increase the Recall but the Precision is decreased. Here, we decided to obtain a higher Precision selecting the most appropriate features using the Gain Ratio.

The second experiment is related to the *Owner Driver identification*, i.e., does the instance belong to the vehicle's owner? In this case, we compared SR only with M since K and G did not calculate the owner driver identification. As stated by the authors of M, they use the same feature set for both the multi-driver and owner driver identification.

The third experiment that we propose is related to the multi-driver identification on the Ψ dataset. This contains data from 14 drivers. Figure 4(b) shows that drivers' instances are not equally distributed. For example, Driver 1 has the highest number of instances with 13617 samples, whereas Driver 10 has the lowest number with only 7 instances. This may depend on the fact that some users drive frequently whereas other users rarely. However, 7 instances are not enough to build a model for the Driver 10. So, we decided to exclude Driver 10 instances in our experiments to not alter the final result. Also, many instances contain empty values because of errors on gathering data. Instances are recorded every 7 seconds.

Compared to the Θ dataset, Ψ contains by default 32 features. Nevertheless, five of these features are timestamp related and are *minute*, *hour*, *day of the week*, *month*, *year*. Other features, such as, *model*, *car_year*, are removed since they do not give any useful information about the user driving style. The dataset also contains *engine_runtime* from which we extract *engine_runtime_minute*.

In this experiment, we used the GR method for features selection. Starting from pruned Ψ dataset, we evaluated SR. As previously stated, we know that there are no other research works that use this dataset. So, we had only the possibility to replicate the best solution proposed by M.

Table II(c) shows that Secure Routine with feature selection achieves the best result both for Precision and Recall.

To conclude the evaluation, last experiment focused on the owner driver identification. Table II(d) indicates that SR with features selection has the best performance when compared with M. SR obtained an average precision of 99,6%, which means that for 8 drivers SR established a perfect Precision whereas M achieved this Precision only for 4 drivers with an average Precision of 95,1%. Regarding the Recall, SR largely

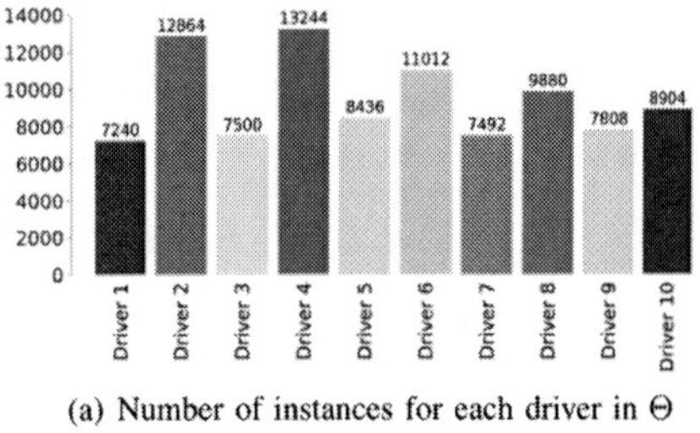

(a) Number of instances for each driver in Θ

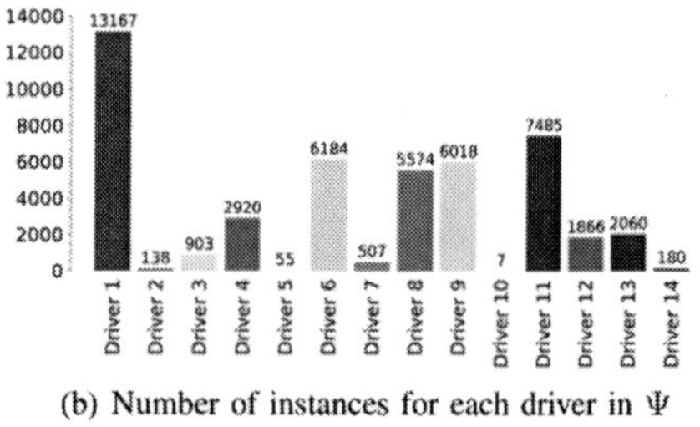

(b) Number of instances for each driver in Ψ

Figure 4. Driver distributions on the datasets.

TABLE I. COMPARING SR USING J48 AND RANDOM FOREST OVER THE MULTI-DRIVER IDENTIFICATION PROBLEM.

J48				Random Forest			
All features		Feature selection		All features		Feature selection	
Precision	Recall	Precision	Recall	Precision	Recall	Precision	Recall
99,2%	99,2%	99,3%	99,3%	99,3%	99,3%	99,6%	99,6%

TABLE II. COMPARISON OF SECURE ROUTINE WITH RELATED WORKS

(a) Comparison of Secure Routine with M, K and G.

Secure Routine		M		K		G	
Precision	Recall	Precision	Recall	Precision	Recall	Precision	Recall
99,6%	99,6%	99,2%	99,2%	N.A.	N.A.	98,8%	98,1%
Accuracy		Accuracy		Accuracy		Accuracy	
99,6%		99,2%		99,6%		N.A.	

(b) Comparison of Secure Routine with M.

Secure Routine		M	
Avg. Precision	Avg. Recall	Avg. Precision	Avg. Recall
99,8%	98,5%	99,3%	99,3%

(c) Comparison of Secure Routine with M for multi-driver identification.

Secure Routine		M	
Precision	Recall	Precision	Recall
99,4%	99,4%	90,4%	89,8%

(d) Comparison of Secure Routine with M for owner identification.

Secure Routine		M	
Avg. Precision	Avg. Recall	Avg. Precision	Avg. Recall
99,6%	98,1%	95,1%	82,9%

outperformed M in percentage and SR achieved a perfect Recall score for one driver, whereas M never obtained a perfect Recall.

VI. CONCLUSION AND FUTURE WORK

In this paper, we introduced for the first time the Secure Routine paradigm to identify the vehicle's owner taking into account the driving style. Also, we presented the algorithm implemented by means of machine learning algorithms and we showed how SR works to identify the driver. Then, we compared SR with other three existing research papers and we evaluated them considering Precision, Accuracy and Recall metrics. Experiments made use of two different datasets. Findings showed that SR obtains the best results compared with the other algorithms considering both experiments regarding the identification of the vehicle's owner and the multi-driver.

As future work, we plan to improve the algorithm of Secure Routine by considering additional features to increase its identification capabilities, i.e., statistical features. We will also improve our *FSParadigm* to enable a better feature selection.

ACKNOWLEDGMENT

This work has been partially supported by the COSCA research project (NGI_TRUST 2nd Open Call 2019002).

REFERENCES

[1] B. Johnston, "Rivervale reveal DVLA data to uncover the most stolen cars in the UK," 02 2020, URL:https://www.rivervaleleasing.co.uk/blog/posts/most-stolen-cars-uk-theft#sthash.iCWvAg4d.dpuf [retrieved: 09, 2020].

[2] Y. Xiong and H. Lin, "Routine based analysis for user classification and location prediction," in 2012 9th International Conference on Ubiquitous Intelligence and Computing and 9th International Conference on Autonomic and Trusted Computing, Sep. 2012, pp. 96–103.

[3] I. Lavie, A. Steiner, and A. Sfard, "Routines we live by: from ritual to exploration," Educational Studies in Mathematics, vol. 101, no. 2, Jun 2019, pp. 153–176, URL:https://doi.org/10.1007/s10649-018-9817-4.

[4] N. Banovic, T. Buzali, F. Chevalier, J. Mankoff, and A. K. Dey, "Modeling and understanding human routine behavior," in Proceedings of the 2016 CHI Conference on Human Factors in Computing Systems, ser. CHI '16. New York, NY, USA: Association for Computing Machinery, 2016, p. 248–260, URL:https://doi.org/10.1145/2858036.2858557.

[5] "The OBDII Home Page", "Obd-ii background," URL:http://www.obdii.com/background.html [retrieved: 09, 2020].

[6] M. Bernardi, M. Cimitile, F. Martinelli, and F. Mercaldo, "Driver and path detection through time-series classification," Journal of Advanced Transportation, vol. 2018, 03 2018, pp. 1–20.

[7] Z. Gao, L. Li, J. Feng, R. Yu, X. Wang, and C. Yin, "Driver identification based on stop-and-go events using naturalistic driving data," in 2018 11th International Symposium on Computational Intelligence and Design (ISCID), vol. 01, Dec 2018, pp. 306–310.

[8] B. Wang, S. Panigrahi, M. Narsude, and A. Mohanty, "Driver identification using vehicle telematics data," in SAE Technical Paper. SAE International, 03 2017. [Online]. Available: https://doi.org/10.4271/2017-01-1372

[9] A. Girma, X. Yan, and A. Homaifar, "Driver identification based on vehicle telematics data using lstm-recurrent neural network," in 2019 IEEE 31st International Conference on Tools with Artificial Intelligence (ICTAI), 2019, pp. 894–902.

[10] HCRL, "Driving dataset," URL:http://ocslab.hksecurity.net/Datasets/driving-dataset [retrieved: 09, 2020].

[11] P. Rettore, "Vehicular traces," 2018, URL:http://www.rettore.com.br/prof/vehicular-trace/ [retrieved: 09, 2020].

[12] B.-I. Kwak, J. Woo, and H. K. Kim, "Know your master: Driver profiling-based anti-theft method," in PST 2016, 12 2016, pp. 211–218.

[13] F. Martinelli, F. Mercaldo, V. Nardone, A. Orlando, and A. Santone, "Who's driving my car? a machine learning based approach to driver identification," 01 2018, pp. 367–372.

[14] F. Martinelli, F. Mercaldo, A. Orlando, V. Nardone, A. Santone, and A. K. Sangaiah, "Human behavior characterization for driving style recognition in vehicle system," Computers & Electrical Engineering, vol. 83, 2020, p. 102504. [Online]. Available: http://www.sciencedirect.com/science/article/pii/S0045790617329531

[15] J. R. Quinlan, C4.5: Programs for Machine Learning. San Francisco, CA, USA: Morgan Kaufmann Publishers Inc., 1993.

[16] C. E. Shannon, "A mathematical theory of communication," Bell System Technical Journal, vol. 27, no. 3, 1948, pp. 379–423. [Online]. Available: https://onlinelibrary.wiley.com/doi/abs/10.1002/j.1538-7305.1948.tb01338.x

[17] A. Cutler, D. R. Cutler, and J. R. Stevens, Random Forests. Boston, MA: Springer US, 2012, pp. 157–175. [Online]. Available: https://doi.org/10.1007/978-1-4419-9326-7_5

[18] L. Breiman, "Random forests," Machine Learning, vol. 45, no. 1, 10 2001, pp. 5–32. [Online]. Available: https://doi.org/10.1023/A:1010933404324

[19] G. M. Hodgson, "The ubiquity of habits and rules," Cambridge Journal of Economics, vol. 21, no. 6, 11 1997, pp. 663–684, URL:https://doi.org/10.1093/oxfordjournals.cje.a013692.

[20] E. Shi, Y. Niu, M. Jakobsson, and R. Chow, "Implicit authentication through learning user behavior," in Information Security, M. Burmester, G. Tsudik, S. Magliveras, and I. Ilić, Eds. Berlin, Heidelberg: Springer Berlin Heidelberg, 2011, pp. 99–113.

[21] "Elm Electronics Inc", "Elm327 obd to rs232 interpreter," 2017, URL:https://www.elmelectronics.com/wp-content/uploads/2016/07/ELM327DS.pdf [retrieved: 09, 2020].

[22] C. A. d. S. Barreto, "OBDdatasets," 2018, URL:https://github.com/cephasax/OBDdatasets/blob/master/masterDegreeResearch/dailyRoutes.csv [retrieved: 09, 2020].

[23] J. Torres, First Contact with Deep Learning, practical introduction with Keras. Watch this space, 7 2018, URL:https://torres.ai/first-contact-deep-learning-practical-introduction-keras/ [retrieved: 09, 2020].

[24] R. Kohavi, "A study of cross-validation and bootstrap for accuracy estimation and model selection," in Proceedings of the 14th International Joint Conference on Artificial Intelligence - Volume 2, ser. IJCAI'95. San Francisco, CA, USA: Morgan Kaufmann Publishers Inc., 1995, p. 1137–1143.

[25] I. Witten, M. Hall, E. Frank, G. Holmes, B. Pfahringer, and P. Reutemann, "The weka data mining software: An update," SIGKDD Explorations, vol. 11, 11 2009, pp. 10–18.

Revisiting Message Generation Strategies for Collective Perception in Connected and Automated Driving

Quentin Delooz and Andreas Festag

Technische Hochschule Ingolstadt / CARISSMA
Ingolstadt, Germany
Email: Quentin.Delooz@carissma.eu
Email: Andreas.Festag@thi.de

Alexey Vinel

Halmstad University
Halmstad, Sweden
Email: alexey.vinel@hh.se

***Abstract*—Collective perception enables vehicles to exchange pre-processed sensor data and is being standardized as a 2nd generation V2X communication service. The European standardization in ETSI foresees the exchange of detected objects and defined a dedicated message type (Collective Perception Message, CPM) with rules to decide when and with which objects the message should be generated, referred to as generation rules. The choice of these rules is not straightforward and influences both channel load and perception quality. For the object inclusion, ETSI currently follows a similar policy as for the generation of Cooperative Awareness Messages (CAM): The objects are filtered based on their dynamics. We regard this approach as conservative. The present paper revisits the generation rules for the CPM and applies two approaches for object inclusion to the CPM – the conservative strategy of ETSI and a more 'greedy' strategy. We assess the performance by discrete-event simulations in a scenario representing a city with realistic vehicle densities and mobility patterns. The simulations take into account the effects imposed by decentralized congestion control. Considering that ETSI currently follows the conservative strategy, we conclude that the application of a greedy strategy improves the perception quality in low-density scenarios.**

Keywords*–*V2X; vehicular communications; collective perception; message generation.

I. INTRODUCTION

Sensor data sharing using vehicle-to-everything (V2X) communications is an effective and low-cost solution to enhance the perception range of a vehicle's sensors. It is the basis for various advanced use cases for connected and automated driving. Recently, the European Telecommunications Standards Institute (ETSI) has completed a study item for sensor data sharing [1], named 'Collective Perception' (CP). CP is based on the periodic exchange of messages with the direct neighbours in communication range. The study item implies important design decisions including the definition of the Collective Perception Message (CPM) and features of the communication protocol towards the future standard.

The collective perception complements other communication services. Specifically, in the European system for V2X communications, the Cooperative Awareness (CA) service enables vehicles to report their position and driving dynamics to others through Cooperative Awareness Messages (CAMs) [2]. Similarly, the CPM carries objects lists, the vehicle's sensor configuration, and other data fields. All message types are transmitted in the bandwidth-limited wireless channels in the 5.9 GHz band allocated for road safety and traffic efficiency applications. Depending on the message frequency and the number of objects included, CPMs can considerably increase the channel load [3]. Decentralized Congestion Control (DCC) limits the overall data rate a vehicle is allowed to transmit over the wireless channel, but introduces additional delays or even drop messages under high channel load [4]. Following the ETSI study item for the CP service [1], the CP protocol defines several mechanisms to reduce the load generated by CPMs. Although DCC achieves the stabilization of the network, it can severely affect the performance of the CP service.

For the selection of objects to include in a CPM, i.e., *inclusion rules*, ETSI has adopted the strategy used for the CAM: By default, a CAM is broadcasted at a rate of 1 Hz. Then, depending on the vehicle dynamics (position, speed, and heading variation over time), the rate increases to up to 10 Hz [2]. In the case of the CPM, the reasoning is the same but applied to each object. Correspondingly to the CAM specification, the Society of Automotive Engineers (SAE) established the Basic Safety Message (BSM) for the DSRC system [5]. Though specified for the same purpose, the default BSM rate is 10 Hz, but independent of the dynamics of the vehicle. In this paper, we apply the BSM approach for the CPM object inclusion, i.e., a detected object will be transmitted at the rate of 10 Hz.

Both CAM and BSM address the trade-off between channel usage and message rate. The CAM generation can be seen as a *conservative* strategy as it uses the channel only when needed even though more transmission resources would be available. In contrast, the BSM generation rules will always send at the maximum rate if the DCC allows it. This strategy can be regarded as a *greedy* approach, which saturates the channel faster. However, in comparison to a conservative approach, it reduces the time between updates for an object.

In this paper, we present the design of the ETSI collective perception service [1] as a decomposition into components for message sending rules, object inclusion, and redundancy mitigation. Inspired by the BSM generation rules, we compare the 'conservative' strategy currently defined by ETSI for object inclusion rules with a 'greedy' approach. The evaluation relies on simulations using the OMNeT++- based ARTERY framework [6] and LIMERIC for DCC [7]. We consider a realistic scenario with urban, suburban, and highway traffic (*LuST* [8]) to evaluate the performance of both approaches.

The remainder of this paper is organized as follows: After reviewing existing work in Section II, we give an overview of ETSI collective perception in Section III and provide technical background on DCC and LIMERIC in Section IV. Section V describes our simulation environment and parameters used to assess the CP performance. Sections VI and VII provide an analysis of the obtained results and conclude the paper.

II. Related Work

Thandavarayan et al. [9] analyse two different policies, which define the object inclusion and message sending rules for the CP service. The *fixed* policy includes all the detected objects and CPMs are generated at a fixed rate. The *dynamic* policy filters the objects based on their dynamics, similarly to the sending rules of CAMs [2]. Additionally, if no object has to be transmitted, the generation of CPMs is omitted. The authors compare both policies in a highway scenario with different vehicle densities, all generating CPMs, but without considering DCC. Garlich et al. [10] analyse the same policies as in [9], but take into account DCC (a reactive approach, see Section IV), message sending rules and different channel configurations. The paper applies two different scenarios: a realistic (*LuST*) and an artificial one ('spider'). The authors of the two papers greatly contributed to the standardization process of the CP service.

Compared to our paper, [9] and [10] do not make a clear distinction between inclusion and sending rules. Specifically, the *fixed* and *dynamic* policies combine different inclusion rules with different sending rules, making the comparison hard to interpret. Additionally, the authors focus on the corner case where all vehicles send V2X messages; even considering a fast-growing rate of V2X-equipped vehicles, this should not happen before years. In the present paper, we make a clearly separate inclusion and sending rules. We analyse in deep the *LuST* scenario and focus on cases where the number of vehicles able to send CPM is low. Additionally, we consider LIMERIC [7], an adaptive DCC approach, which is more permissive compared to the reactive approach in [10].

III. Overview of Collective Perception Standardized in ETSI

Based on the study item of ETSI about Collective Perception in [1], we decompose the CP service into components for triggering, inclusion, redundancy, and sending rules, which are periodically checked and subsequently executed (see Figure 1). We note that the component names do not correspond directly to the terms in [1], but our proposal eases the understanding of the mechanisms and their relationship. In addition, we do not consider the segmentation of CPMs as in [1]. Instead, if the size of the CPM is larger than the maximum message size of 1,100 B, we randomly remove objects from the message until the maximum message size is reached. Object removal in our scenario occurs rarely and can therefore be neglected. In the following, we explain each component.

A. Checking time

The checking time determines the frequency with which the rules are periodically inspected. It can be regarded as a sleeping time of the algorithm, i.e., the time duration in which a CPM cannot be generated. The value should be less than or equal to the minimum interval between two consecutive CPMs, i.e., 100 ms. Though [1] does not define a checking time, we can assume the same value as specified for the CA service [2] (see Section III-B). In addition, most of the existing research publications use a value of 100 ms.

B. Triggering rules

These rules define the time to wait between the generation of two consecutive CPMs. In [1], the lower and upper bound of the CPM transmission interval time is set to 100 ms and 1 s, respectively. DCC regulates the transmission rate of the CP service between these bounds. If DCC allows, CP triggers the generation of a CPM and set its content with the rules defined by the next components in Figure 1. Both *checking* and *triggering* rules are independent of the *conservative* and *greedy* policy.

C. Locally perceived environment

This component subsumes the pre-processed sensor data as a set of detected, tracked, and classified objects in a vehicle. In general, depending on its technical characteristics, each sensor type represents an object differently. However, using the CP service, the objects are represented in a standardized format, i.e., by their descriptions including position and speed relative to a reference position of the sending vehicle.

D. Inclusion rules

This component filters less relevant objects and these with a confidence level below a pre-defined threshold. The ETSI study item on collective perception [1] defines relevance criteria based on the objects' dynamics, type, and last transmission time. The criteria for object dynamics rely on the CAM generation rules [2], i.e., on the object's difference in position, speed, and heading since the last object inclusion. For the confidence level and threshold, we note that [1] does not define the parameter values.

Figure 2 depicts the decision tree that is executed to decide if an object should be included in the generated CPM, or not. We stress that these inclusion rules rely only on the perception of the sending vehicles, i.e., the objects received via V2X communication are not taken into account. Furthermore, the inclusion rules are static and are applied irrespective of the channel load or the vehicle's driving situation.

In the present paper, we compare the inclusion rules defined in [1], i.e., the *ETSI rules*, with an approach where all objects detected are included in the generated CPM, i.e., *no-filtering*. In analogy to the channel usage vs. message generation trade-off discussed in Section I, the ETSI inclusion rules are seen as *conservative* and non-filtering as *greedy*.

E. Redundancy mitigation rules

A redundant transmission occurs when the same object is received multiple times from different senders. The redundancy mitigation rules omit the transmission of objects which were already received. In [1], these rules are only applied if the channel load is larger than a (still undefined) threshold. Additionally, [1] proposes several strategies. For example, the frequency-based approach omits locally perceived objects from the new CPM if a certain number of previously sent CPMs in a given time window already included information about the same objects. As proposed by [3], some of these rules could also be considered as *inclusion rules*.

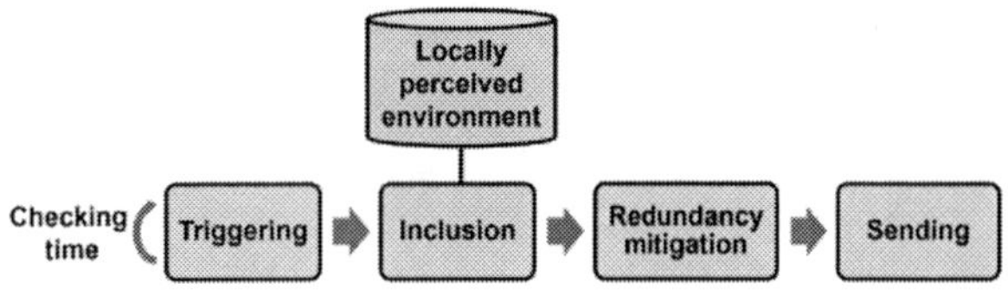

Figure 1. Components for the message generation in ETSI collective perception derived from [1] (without segmentation).

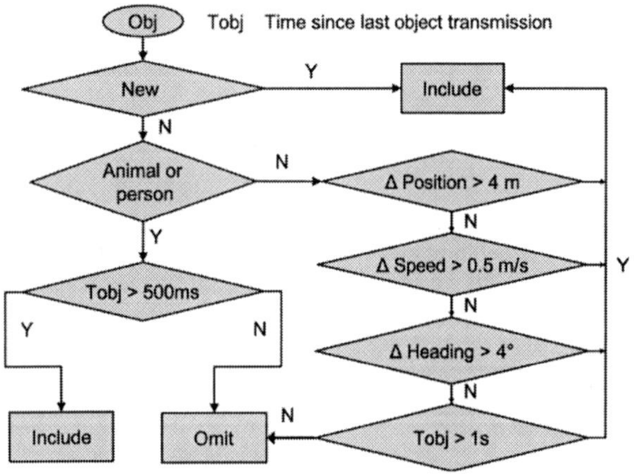

Figure 2. Rules for object inclusion as defined in [1], which corresponds to the *conservative* strategy in this paper.

A typical scenario where the redundancy mitigation rules would be effective is a road intersection. In that case, potentially many vehicles detect the same object, e.g., a pedestrian, and start transmitting information about it. The added information brought by each vehicle would be small in comparison to the 'cost' of its transmission, especially in areas with a dense number of vehicles.

F. Sending rules

The sending rules represent the last decision point to decide if the generated CPM should be sent to the lower levels of the protocol stack. Following [1], at least one of these three conditions should be respected for the CPM to be generated: *(i)* there is at least one object to send, *(ii)* the last CPM with sensor information and *(iii)* the duration since the generation of the last CPM is at least 1 second.

In [10], the authors compare the non-filtering and the ETSI inclusion rules. However, the authors apply different sending rules for the inclusion strategy: With the ETSI inclusion rules, a CPM is generated if objects are present to be transmitted. For the non-filtering approach, a fixed rate of 10 Hz is applied, independently whether objects have to be transmitted. Therefore, it is hard to analyse the effect of the inclusion rules and the sending rules separately, especially if the sensors' parameters do not allow vehicles to detect always at least one object. In the present study, CPMs will only be generated if there is information, such as objects or sensors data, to send, and independently of the used inclusion rules.

IV. Decentralized Congestion Control (DCC)

In the following, we provide background information on DCC in general in Section IV-A and specifically on LIMERIC in Section IV-B.

A. General

DCC is a set of mechanisms in the V2X protocol stack that ensure the stability of the network and fairness in resource usage among network nodes. Its principal function is to measure the channel load (channel busy ratio, CBR) and to control the data that a station generates. DCC is standardized by ETSI in several standards.

DCC is a cross-layer functionality with interacting entities at different layers. The access layer functionality [11] provides traffic shaping for the injected packets. Practically, it implements a 'gatekeeper' that realizes a First-In-First-Out (FIFO) queuing system for each channel. A gatekeeper has multiple queues for the packets to be sent and a single server, which dispatches always the non-empty queue with the highest priority (simple priority queue). When a packet enters the gatekeeper, and the queue is not full, DCC allows the transmission of the packet and sets its transmission parameters. If the queue is full or the lifetime of the packet expires during the waiting time in the queue, the packet is discarded.

To determine when a packet can be transmitted to the MAC layer, ETSI standardized two types of strategies for the gatekeeper [12]: reactive and adaptive. Both strategies respect the DCC requirements specified in [13]:

- $0 < T_{on} < 4ms$: T_{on} is the maximum duration of a packet transmission.
- duty cycle $<= 3\%$: it means that a station can occupy at most 3%, i.e., 30 ms, of channel time.
- $T_{off} >= 25ms$: T_{off} is the duration before the gatekeeper re-opens after the transmission of packet and allows a new packet to be transmitted. In other words, the maximum packet transmission frequency is 40 Hz.
- if $CBR >= 0.62$, $T_{off} >= 1{,}000ms$.

The reactive approach defines a set of states for which values of the T_{off} time are assigned to specific CBR thresholds. The higher the measured CBR, the longer a station needs to wait between two consecutive transmissions. [12] proposes two sets of states, each one depending on the maximum allowed transmission time. Effectively, the reactive approach sets a predefined rate based on the measured CBR. In contrast, an adaptive method shares the channel resources between the stations in communication range such that the CBR converges to a predefined maximum value. The LInear MEssage rate Integrated Control (LIMERIC) algorithm [7] meets the ETSI requirements for the adaptive DCC approach and is used in the simulation of the present paper.

B. LIMERIC

Instead of directly adapting the transmission rate, LIMERIC adjusts the duty cycle δ every 200 ms. The duty cycle is the allowed ratio of the transmitter total "on" time relative to 1 s. [12] defines the algorithm to adapt δ depending on the observed CBR. [14] provides some insights about the reason behind the chosen LIMERIC parameters and proposes a dual-α approach to improve LIMERIC's convergence time

TABLE I. SUMMARY OF LIMERIC'S PARAMETERS

Parameter	Description	Values
α_{low}	Convergence parameter	0.016
α_{high}	Convergence parameter	0.1
th	To choose between the α	0.00001
α	Convergence parameter	as in [14]
β	Convergence parameter	0.0012
CBR_{target}	Convergence point	0.68
δ_{max}	Max allowed duty cycle	0.03
δ_{min}	Min allowed duty cycle	0.0006
δ_{init}	Initial δ	0.0153
G^{+}_{max}	Upper born used to update δ	0.0005
G^{-}_{max}	Lower born used to update δ	-0.00025
T_{CBR}	Interval for CBR value update	100 ms

Figure 3. Topology of Luxembourg in *LuST*.

and fairness during transition phases. We decided to use this modification. Table I summarizes the LIMERIC parameters used in our simulations.

From the allowed duty cycle determined by LIMERIC, [12] derives T_{off} to enforce the rate by

$$T_{off} = \min(\max(\frac{Ton_{pp}}{\delta}, 25\,\text{ms}), 1\,\text{s}) \qquad (1)$$

with Ton_{pp} being the transmission time of the last transmitted packet. We note that the reactive strategy considers only the CBR and makes some simple assumptions for the packet size. In contrast, the adaptive strategy takes into account the size of the transmitted packet to enforce the allowed duty cycle.

V. SIMULATION ENVIRONMENT

This section presents the used simulation framework, the V2X services deployed and their respective message formats, and how the vehicles are equipped in the simulations.

A. Simulation framework

For the evaluation of the CPM generation strategies, we used the discrete-event simulator ARTERY [6] to model the V2X communications following ETSI standards. *ARTERY* relies on VANETZA, INET and OMNeT++ (v5.4.1), and implements the V2X protocol stack based on ITS-G5 (see [6] for details). To model the traffic and mobility of the vehicles, we used the microscopic road traffic simulator *SUMO* (v1.0.1) [15] with the popular Luxembourg scenario, a.k.a. *LuST* [8]. Figure 3 shows the topology of the SUMO map for the Luxembourg scenario (see [3] for the distribution of vehicles). We note that the LuST scenario was validated with real mobility data for SUMO version 0.26. Since we have used a newer version of SUMO, the traffic mobility model cannot be regarded as formally validated but still represents a realistic scenario.

Each simulation run is executed for a duration of 13 s with 10 s of warmup. The warmup phase gives time to LIMERIC to converge to the desired δ for each vehicle.

For the *LuST* scenario, we have chosen a snapshot at 8 a.m. This corresponds to a rush-hour with around 5,000 vehicles in the simulated environment. Within the scenario, we selected three distinct areas: urban, suburban, and highways, respectively represented in Figure 3 by the blue, orange, and black squares. In terms of vehicle density, the *urban* area will face the highest density and the *suburban* the lowest. Table II shows the vehicle dynamics depending on the area; the dominant dynamic parameter are marked in bold. For example, for the *highway* area, the dominant parameter is Δ position, i.e., the vehicle speed, because it triggers the generation of a CAM first. Following the vehicle dynamics parameters for CAMs [2], the theoretically resulting CAM transmission rate is presented in the last row of Table II. For the urban area, the two parameters Δ position and Δ speed result in an almost equal CAM transmission rate. Therefore, we indicated both as dominant in Table II.

B. V2X services and CPM format

Both CA and CP services are enabled. The CA service operates on the Control Channel (CCH) and the CP service on the Service Channel 1 (SCH1) of the 5.9 GHz frequency band. We consider that the vehicles can receive and send at the same time on both channels and that there is no interference between them. The fading model used is the one integrated into ARTERY called *VanetNakagamiFading* [16].

The CAM and the CPM formats rely on [1] and [2], respectively. Specifically, the CPM consists of an ITS PDU header and several containers, including containers for management and station data containing information about the sender such as position, heading and velocity, 0 to 127 Sensor Information Containers (SICs), and 0 to 127 Perceived Object Containers (POCs). Using default values, the size of a SIC varies from 11 to 88 bytes and a POC from 20 to 46 bytes. The CPM format is specified in ASN.1 and encoded by the Unaligned Packed Encoding Rules in ASN.1 as specified by ETSI (see Section 6.8.3 and Annex A of [1]). We have used the ASN.1 open-source compiler *asn1c*.

TABLE II. AVERAGE CHANGES OF THE VEHICLE DYNAMICS DURING A 100 ms TIME INTERVAL IN THE *LuST* SCENARIO

	All	Urban	Suburban	Highway
Δ position [m]	**1.415**	**0.777**	**0.996**	**2.47**
Δ speed [m/s]	0.071	**0.097**	0.102	0.042
Δ heading [°]	0.006	0.006	0.01	0.007
CAM frequency (Hz)	3.54	1.94	2.49	6.175

TABLE III. SUMMARY OF THE SIMULATION PARAMETERS

Parameter	Values
Protocol stack	ITS-G5
Frequency band	5.9 GHz
Channel number (IEEE numbering scheme)	SCH1 (176) for CP service CCH (180) for CA service
Channel model	*VanetNakagamiFading*
DCC	LIMERIC
Inclusion rules	{Etsi, No-filtering}
Scenarios	*LuST*
PVE	{10, .., 90, 100}
Time of simulation	8 a.m.
Number of vehicles	≈ 5,000
Simulation time	13 s (incl. 10 s of warmup)
Number of repetitions	2
Vehicle sensor equipment	{60 & 174 m, +/-10 & +/-45°} {150 m, 360°}

C. *Vehicle equipment and object detection*

It is assumed that with the increasing deployment of C-ITS, the ratio of vehicles equipped with V2X technologies will grow over time. The larger the ratio, the higher gets the generated data load on the channel. To analyse the impact of the V2X equipment rate on the performance of the filtering approaches, we varied the V2X equipment rate (PVE = Percentage of Vehicles Equipped) and used the values PVE = {10 20, ..., 100} %.

For object detection, the vehicles have local sensors mounted on them and we used two different configurations. In the first one, each vehicle has two radars with respectively a range of 60 and 174 m, and a field of view of +/– 45 ° and +/– 10 °, respectively. Both radars are located in front of the vehicle and are facing ahead. In the second configuration, each vehicle is equipped with a radar with a range of 150 m, and a FOV of 360 °. The first configuration simulates the early development of sensor perception. The second one grossly reproduces the future perception capabilities of vehicles.

The method to detect objects is the same as explained in [17]. In brief, each ITS-S mounted with sensors detects an object if one of the four corners of the object is in the line of sight of one of the sensors. The information retrieved from the perception is idealistic, i.e., all object attributes are always available and no errors in object detection occur.

VI. PERFORMANCE EVALUATION

The evaluation compares the performance of the conservative and the greedy strategy for object inclusion in the CP service for different values of the PVE. In the evaluation, we also vary the area type (urban U and highway H) and the sensor vehicle equipment (Field of view of {+/-10 & +/-45°} and 360°). For example, the simulation *U-CPM-conservative (360)* corresponds to the *conservative* strategy for object inclusion in the *urban* scenario and with a vehicle sensor configuration for an FOV of *360°*. We collected different metrics to assess distinct aspects of the *CPM-conservative* and *CPM-greedy* inclusion rules. For readability reasons, we only include results for the urban and highways areas, which cover most of the interesting points to discuss.

A. *Network-related metrics*

The *Channel Busy Ratio* (CBR) gives a measure of the channel occupancy. The *Packet Error Rate* (PER) is the rate of unsuccessfully decoded messages on the number of received ones. The higher the CBR, the larger the PER is likely to be. Figure 4a shows the CBR on the SCH1, i.e., the CBR obtained with the CP Service, for the urban (U) and highways (H) areas in the *LuST* scenario. In general, the CBR for the urban area is higher than for the highway one. The same applies to Figure 4b showing the PER obtained for the same scenarios. The highest CBR is observed with the *greedy* approach and a sensor with a FoV of 360°. With this configuration, the PER is around 25 % at PVE=100 %. Still, the CBR does not reach the targeted CBR defined by LIMERIC, and the channel is not saturated. The ETSI configuration does not generate sufficient data for a CBR higher than 0.3. The maximum average observed PER obtained for ETSI is around 25 % with the *CPM-greedy (360)* configuration in the urban scenario.

B. *Application-related metrics*

Figure 5 shows the results obtained for the *Number of Objects Detected* (NOD) and the *Time Between Update* (TBU) metrics. The NOD metric represents how many objects a vehicle was aware of during the last second. The TBU metric expresses the average time between two consecutive updates of the same object. Both CAMs and CPMs contribute to these metrics. In the urban scenario and with the 360° sensor, the *greedy* approach provides the highest number of NOD for any PVE. Interestingly, with this configuration, when the PVE reaches 50%, the NOD does not vary significantly anymore. We could question the necessity to have more vehicles sending CPMs if we can assert that CPM transmitter distribution is uniform. Additionally, the *U-CPM-conservative (360)* and the *U-CPM-greedy* have comparable results. The urban area with ETSI inclusion rules has the highest TBU. However, the updating rate is always less than if only the CA service would be used (see Table II). The smallest TBU, around 50 ms, is obtained in the highway area, a 360° sensor, and the *greedy* inclusion rules. The difference between *greedy* and *conservative* is more important in the urban than for the highway. This is expected since in highway scenarios, vehicles exhibit higher dynamics than in the urban area (see Table II).

C. *Ratio of Resource Used (RRU)*

Figure 6 shows the average *Ratio of Resources Used* (RRU) by a CPM. Following (1), the RRU is defined by Ton_{pp}/δ. For example, if $\delta = 0.001$ and the transmission time of a CPM is 200µs, then the RRU taken by this CPM will be 0.2. This means that the transmitted CPM takes 20% in a 1 s time duration of the channel access time determined by LIMERIC. If the RRU is lower than 0.1, DCC always allows the CP service to generate CPMs at the maximum rate of 10 CPM/s. Except for the *U-CPM-greedy (360)* configuration, the RRU is in average constant, independently of the PVE. This can be explained by two reasons: the average size of CPM for each configuration is the same, independently of the PVE. The second reason is the convergence time of LIMERIC when the parameter δ increases. Indeed, even with 10 s of warmup, corresponding to 50 updates of δ, its highest possible value would be around 0.023 from δ_{init}. This is not a problem as the resulting T_{off} is always shorter than 100 ms. With the dual-alpha approach [14], δ converges faster when decreasing.

For the *U-CPM-greedy (360)*, the δ decreases enough to observe the RRU increasing. It means that with more

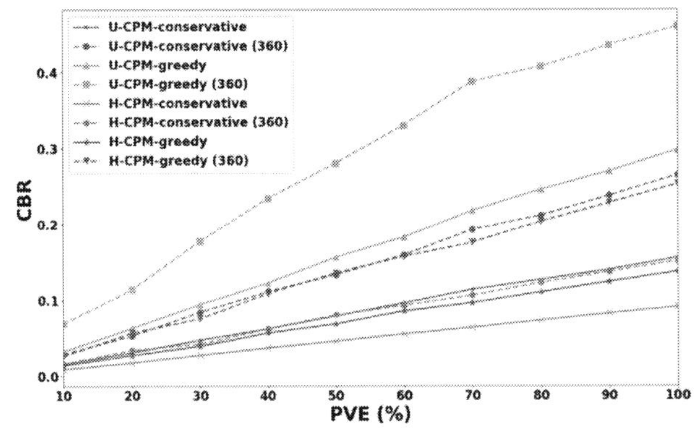

(a) Average Channel Busy Ratio (CBR)

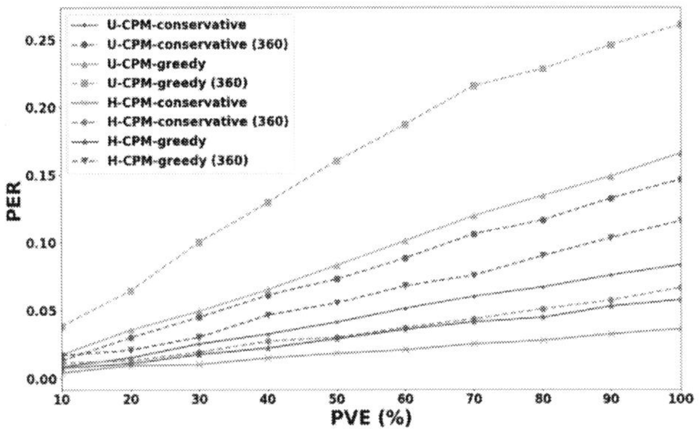

(b) Average Packet Error Rate (PER)

Figure 4. Network-related metrics for different values of percentage of vehicles equipped (PVE) with V2X capabilities in the *LuST* scenario.

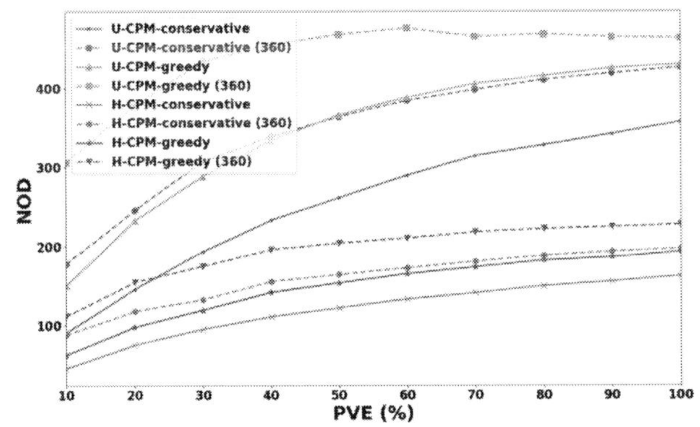

(a) Average Number of Objects Detected (NOD)

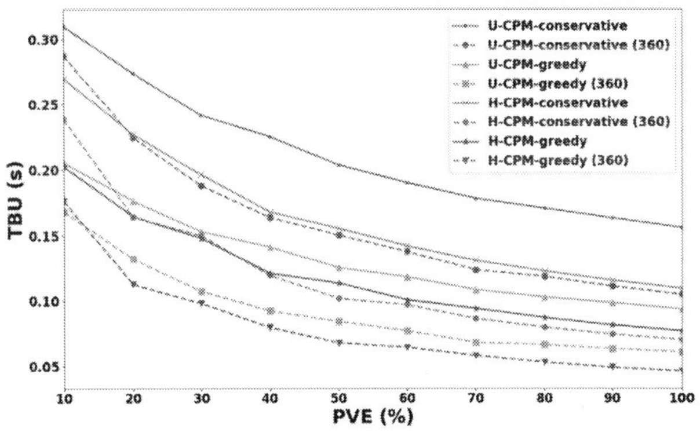

(b) Average Time Between Updates (TBU)

Figure 5. Applications-related metrics for different PVE values in the *LuST* scenario.

transmitting vehicles and objects to detect, LIMERIC would only start reducing the transmission rate of the vehicles. Only in the configuration *U-CPM-greedy (360)*, the RRU increases with the PVE. Still, in average the RRU remains under 0.1. Therefore, the CP service can generate 10 CPM/s.

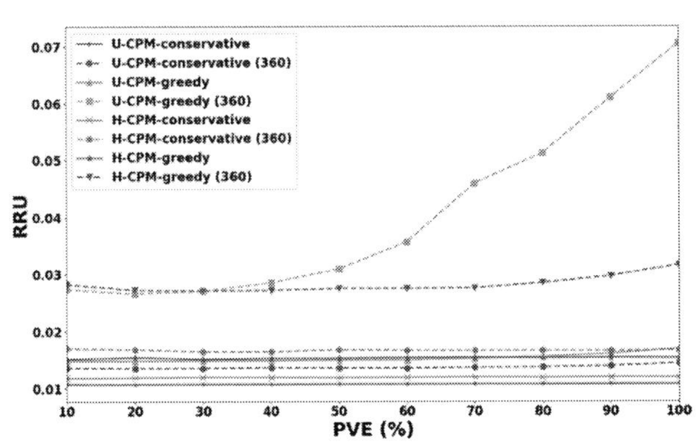

Figure 6. Average Ratio Resource Used (RRU).

D. Discussion

Even if not all kinds of objects, such as pedestrians or obstacles, are present in the simulations, both scenarios provide insights about the trade-off between channel load and perception quality. In the *LuST* scenario, the CP service is not able to saturate the channel even without filtering and with a PVE of 100 %. Notably, the *conservative* inclusion rules underutilize the channel resources, while largely available, at low PVE and in areas with a small density of vehicles. Compared to *conservative*, the *greedy* approach provides always a better perception quality while not saturating the channel.

VII. Conclusion and Future Work

In this paper, we addressed the CP service as currently defined by ETSI and presented it as a decomposed system with message sending rules, object inclusion, and redundancy mitigation. For the object inclusion rules, we analysed two strategies, i.e., *conservative* and *greedy*. In comparison, the *greedy* strategy, which does not filter objects, provides a reduced time between updates and a higher number of perceived objects at the cost of higher channel usage. Still, in the considered scenario the target CBR of LIMERIC is never reached. Also,

DCC does not have to reduce actively the message rate. While both *greedy* and *conservative* approach differently consider the trade-off between the perception quality and channel usage, at a low percentage of equipped vehicles and in areas with a small density of vehicles, the *greedy* approach allows for a higher channel utilization and for better performance.

In our future work, we will study the improvement of the CP service when the inclusion rules dynamically adapt to the channel load. This approach smoothly combines the *greedy* and *conservative* approach: When the channel load is low, fewer objects are filtered and, the quality of perception is increased. When the channel is close to saturating, we switch to the conservative approach and filter more objects. We will also consider other filtering approaches, such as redundancy mitigation rules, to reach the same goal.

ACKNOWLEDGMENT

This work was supported by the German Research Foundation DFG within the priority program Cooperative Interacting Cars (CoInCar) (SPP 1835) and in part by the Swedish Knowledge Foundation within SafeSmart "Safety of Connected Intelligent Vehicles in Smart Cities" Synergy Project (2019–2023). The authors would like to thank H. Güenther (Volkswagen), R. Riebl (Technical University Ingolstadt), K. Garlich and L. Wolf (Technical University Braunschweig) for valuable discussions.

REFERENCES

[1] ETSI, "Intelligent Transport Systems (ITS); Vehicular Communications; Basic Set of Applications; Analysis of the Collective Perception Service (CPS); Release 2," Dec. 2019, ETSI TR 103 562 V2.1.1.

[2] ETSI, "Intelligent Transport Systems (ITS); Vehicular Communications; Basic Set of Applications; Part 2: Specification of Cooperative Awareness Basic Service," Apr. 2019, ETSI EN 302 637-2 v.1.4.1.

[3] Q. Delooz and A. Festag, "Network load adaptation for collective perception in V2X communications," in 2019 IEEE International Conference on Connected Vehicles and Expo (ICCVE), 2019, pp. 1–6.

[4] S. Kühlmorgen et al., "Evaluation of congestion-enabled forwarding with mixed data traffic in vehicular communications," IEEE Transactions on Intelligent Transportation Systems, vol. 21, no. 1, 2020, pp. 233–247.

[5] J. B. Kenney, "Dedicated Short-Range Communications (DSRC) standards in the United States," Proc. IEEE, vol. 99, no. 7, Jul. 2011, pp. 1162–1182.

[6] R. Riebl, H. Günther, C. Facchi, and L. Wolf, "Artery: Extending Veins for VANET applications," in 2015 International Conference on Models and Technologies for Intelligent Transportation Systems (MT-ITS), pp. 450–456.

[7] G. Bansal, J. B. Kenney, and C. E. Rohrs, "LIMERIC: A linear adaptive message rate algorithm for DSRC congestion control," IEEE Trans. Veh. Technol., vol. 62, no. 9, Nov. 2013, pp. 4182–4197.

[8] L. Codeca, R. Frank, S. Faye, and T. Engel, "Luxembourg SUMO traffic (LuST) scenario: Traffic demand evaluation," IEEE Intell. Transp. Syst. Mag., vol. 9, no. 2, 2017, pp. 52–63.

[9] G. Thandavarayan, M. Sepulcre, and J. Gozalvez, "Analysis of message generation rules for collective perception in connected and automated driving," in 2019 IEEE Intelligent Vehicles Symposium (IV), 2019, pp. 134–139.

[10] K. Garlichs, H. Günther, and L. C. Wolf, "Generation rules for the collective perception service," in 2019 IEEE Vehicular Networking Conference (VNC), 2019, pp. 1–8.

[11] ETSI, "Decentralized Congestion Control Mechanisms for Intelligent Transport Systems operating in the 5 GHz range; Access layer part," Apr. 2018, ETSI TS 102 687 V1.2.1.

[12] ETSI, "Intelligent Transport Systems (ITS); Decentralized Congestion Control Mechanisms for Intelligent Transport Systems operating in the 5 GHz range; Access layer part," Apr. 2018, ETSI TS 102 687 V1.2.1.

[13] ETSI, "Intelligent Transport Systems (ITS); Radiocommunications equipment operating in the 5 855 MHz to 5 925 MHz frequency band; Harmonised Standard covering the essential requirements of article 3.2 of Directive 2014/53/EU," Feb. 2017, ETSI EN 302 571 V2.1.1.

[14] I. Soto, O. Amador, M. Urueña, and M. Calderon, "Strengths and weaknesses of the ETSI adaptive DCC algorithm: A proposal for improvement," IEEE Commun. Lett., vol. 23, no. 5, May 2019, pp. 802–805.

[15] P. A. Lopez et al., "Microscopic traffic simulation using SUMO," in 2018 21st International Conference on Intelligent Transportation Systems (ITSC), 2018, pp. 2575–2582.

[16] L. Cheng, B. E. Henty, D. D. Stancil, F. Bai, and P. Mudalige, "Mobile vehicle-to-vehicle narrow-band channel measurement and characterization of the 5.9 GHz dedicated short range communication (DSRC) frequency band," IEEE J. Sel. Areas Commun., vol. 25, no. 8, Oct. 2007, pp. 1501–1516.

[17] H. Günther, O. Trauer, and L. Wolf, "The potential of collective perception in vehicular ad-hoc networks," in 2015 14th International Conference on ITS Telecommunications (ITST), 2015, pp. 1–5.

Proposal of Guidance Method in Car Navigation Systems

Akimasa Suuki, Yoshitoshi Murata, Nobuyoshi Sato
Faculty of Software and Information Science
Iwate Prefectural University
Takizawa, Japan
e-mail: {suzuki_a, y-murata, nobu-s}@iwate-pu.ac.jp

Ikuya Soma
Business Strategy Department
NTT Data Inc.
Tokyo, Japan
e-mail: Ikuya.Soma@nttdata.com

***Abstract*—Automotive navigation systems have become increasingly popular. Web-based map applications, such as Google Maps, that are available on smartphones are commonly used in automotive navigation systems. Such systems help drivers navigate through unfamiliar regions. However, drivers may occasionally misjudge or make an incorrect turn in response to voice directions such as "In XX meters, turn right". This possibly happens because of the difference between perceived and actual distance. This study aims to examine the differences in the perceived and actual values of not only distance but also elapsed time. It then proposes a guidance method that instructs drivers to turn on their right or left blinker before approaching a turn. The method reduces the likelihood of drivers missing turns and does not bother drivers with repeated instructions. In addition, the method ensures that the cars behind are aware of the driver's intended actions as the blinkers are turned on prior to making the turn; thus, it is superior in terms of driving safety.**

Keywords-Car navigation; driving safety; guidance method; voice instruction; perceptual distance.

I. INTRODUCTION

Automotive navigation systems have become increasingly popular. Web-based map applications, such as Google Maps [1], that can be accessed on mobile devices (e.g., smartphones) are a commonly used automotive navigation system, helping drivers navigate through unfamiliar regions. However, despite voice instructions such as "in XX meters, turn right", drivers may occasionally miss a turn, go in the wrong direction, or make a last-minute turn, which can be dangerous. This is possibly because drivers miscalculate the indicated distance.

This research examines the difference between the perceived and actual values of not only distance but also elapsed time. The experiments show that most participants could accurately perceive distance within 100 m, although this accuracy rapidly decreased when the distance increased to more than 100 m. Thus, we conclude that re-instructing drivers within 100 m of a turn helps them accurately perceive distance.

We firstly considered that the voice countdown method based on conventional visual count down bar [2] was the most effective in accurately perceiving distance. In this paper, we evaluated the countdown method. However, the method tended to offer repeated instructions that can prove bothersome to a driver. Thus, we proposed an instruction method in which drivers are asked to turn on their right or left blinker as they are nearing a turn. The experimental evaluations reveal that the method prevents drivers from making wrong turns and helps them drive safely, although the accuracy of perceived distance is marginally lower than that observed in the countdown method. The method does not bother a driver with repeated instructions. Moreover, it is superior in terms of driving safety because the cars behind become aware of the driver's intended actions when the latter turns on the blinker.

The remainder of this paper is organized as follows. Section II discusses related works. Section III evaluates the accuracy of perceived distance and time. Section IV analyzes the accuracy of selecting an intersection. Section V presents the novel instruction method. Section VI tests the proposed instruction method by conducting related experiments. Section VII compares the proposed method with existing ones. Section VIII offers concluding remarks and suggestions for future research.

II. RELATED WORKS

Automotive navigation systems have three main tasks: positioning, routing, and navigation (guidance). This study focuses on navigation or guidance methods. Guidance methods generally include the display of instructions on a road map in a navigation system, an information display on a windshield, and/or voice instructions.

However, studies have shown that the continuous need to look at the navigation system's display for information can be distracting and dangerous [3]. Therefore, most navigation systems are a combination of a road map display and voice instructions. Guidance information generally includes distance to the destination and landmarks that can help drivers locate the destination [4]. It is easy to miss a turn in response to distance-based instructions such as "In XX meters, turn right/left." Thus, manufactures of car navigation systems have upgraded guidance methods by, for example, integrating a 3D map that improves identifiability of mapped roads in reality. Figure 1 is an image of Panasonic's car navigation system with a 3D map, Strada CN-F1XVD [5]. Augmented reality (AR) technologies have also contributed to improving identifiability in guidance instructions [6]-[8]. Akaho et.al. [6] analyzed the AR methods and comfirmed ease of understanding, safety, and the characteristics of AR-Navi in comparison with conventional method using 3D maps. However, 3D maps and AR technologies do not resolve the problem of drivers having to constantly look at the display for instructions.

The second guidance method, that is, an information display on a windshield, was developed to resolve the abovementioned issue. Figure 2 depicts an AR navigation system and windshield projection unit designed by Pioneer Corporation. The AR navigation system is to the lower left of the image. The in-vehicle camera captures a real-time video of the road and vehicles ahead of the car, and guidance information is overlaid using computer-generated imagery on a live feed using AR technology. The information on the AR navigation system is also projected on a see-through windshield display.

Large and Burnett [9] examine the effects of different types of voice navigation described in [6]. However, to the best of our knowledge, no study explores ways to improve the guidance accuracy of voice navigation systems

III. Perceived Distance and Elapsed Time

Existing automotive navigation systems offer driving instructions such as "in XX meters, turn right" or "turn left at the intersection." However, the difference between perceived and actual distance may cause drivers to misjudge a turn or even steer abruptly, which can be dangerous. This study, therefore, conducts experiments to evaluate differences between the perceived and actual values for distance and elapsed time.

Figure 1. Panasonic's car navigation system, Strada CN-F1XVD [1]

Figure 2. Pioneer Corporation's Cyber Navi [3]

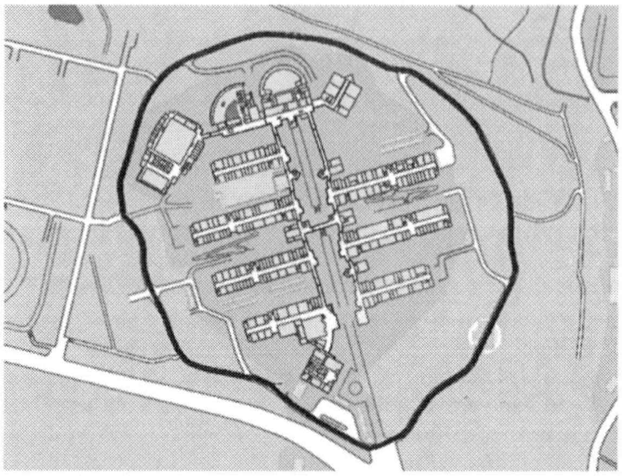

Figure 3. Test course in the university

A. Experiment for halting vehicles

A total of 20 students with a driving license were asked to drive a Toyota Noah installed with a navigation system on a circuit road in Iwate Prefectural University (see Figure 3). Each participant is informed of the distance and time (in seconds) within which they must halt the car. The distances used in this experiment are 100, 300, and 500 m. Each participant must drive and stop the car thrice for each distance category. The elapsed times are 10, 20, and 30 s, and the participants must drive and stop the car for each elapsed time.

Prior to estimating the perceived values and conducting the experiment, the participants were asked to drive few laps of the test course to ensure they understand the examiner's instructions and to confirm the distance and elapsed time.

B. Experiment results for perceived distance

Table I presents the average distance, average difference between the instructed and perceived distance, and standard deviations. If the instructed distance is longer, the average difference tends to be larger. This difference varies by as much as ten percent among the participants, as indicated by the standard deviations in Table I. In other words, each participant perceives distance differently. This means that additional instructions are needed for distances less than 100 m, and instructions such as "In XX meters, turn right/left" are not effective when the distance is greater than 100 m.

Table I. Estimation results for perceived distance

	100 m	300 m	500 m
Average (m)	138	327	527
Avg. difference (m)	41	82	120
Std. dev. (m)	38.7	91.3	121.6

Figure 4 shows the frequency distribution for each instructed distance. The vertical axis denotes the frequencies employed in the experiments and the horizontal axis indicates the difference between the instructed and travelled distances. The frequency distribution is estimated for iterations for each instructed distance. The results suggest that the data are widely distributed, and the curve does not take the form of a normal distribution. Some participants travelled considerably further than the instructed distance, although no participant travelled significantly less than the instructed distance. These

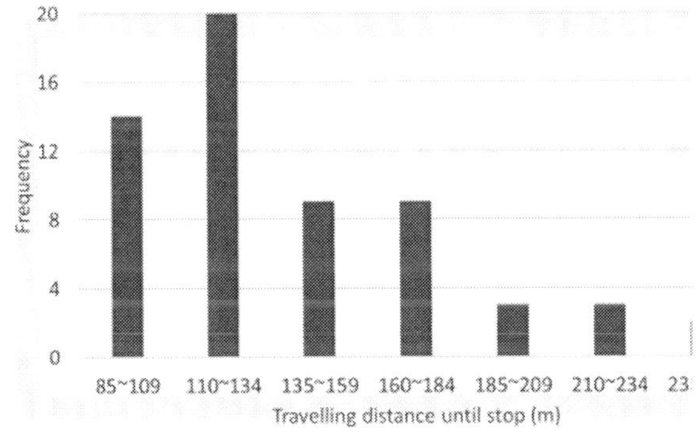

(a) Instruction distance = 100 m

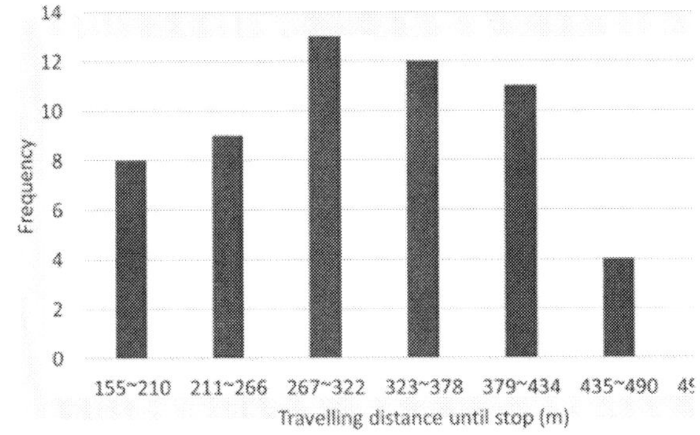

(b) Instruction distance = 300 m

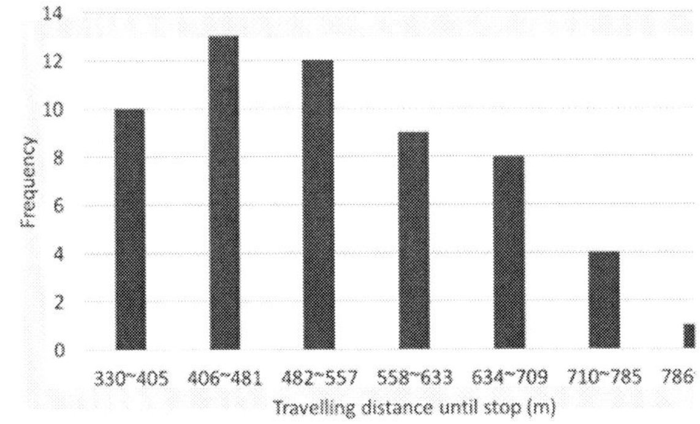

(c) Instruction distance = 300 m

Figure 4. Frequency distribution for an instructed distance

results suggest that re-instructing drivers when they are within 100 m of a turn or destination decreases the likelihood of them going in the wrong direction.

C. Experimental result for perceived elapsed time

Table II presents average distance, average difference between instructed and travelled time, and standard deviations in the context of elapsed time. Figure 5 presents the frequency distribution for each instructed elapsed time. The vertical axis indicates the frequencies applied in the experiments and the horizontal axis denotes the ranges of time elapsed between the instructed time and the vehicle halting. The frequency distribution is estimated from 60 iterations for each elapsed time.

The results reveal significantly small deviations from an instructed time (<15%). When the instructed time is 10 s, the error distance is roughly 20 m at 60 km/hour.

A method that notifies drivers of the elapsed time would be more effective that one with distance instructions. However, it is difficult to estimate elapsed time on urban roads or streets since drivers seldom maintain a consistent speed. Nevertheless, the method can be useful on regional roads such as highways, where drivers are generally expected to travel at a constant pace.

The next section proposes a guidance method that can prove more effective than the distance instruction method.

IV. Novel Guidance Method

The results in the previous section indicate that re-instructing drivers within 100 m of a turn or destination decreases the likelihood of them driving in the wrong direction.

We propose the following alternative guidance methods:

1) A countdown method that provides drivers with voice instructions at various distance intervals (50, 40, 30, 20, 10, and 0 m).
2) A blinker method that instructs drivers to turn on their blinker at, for example, 30 m before the target intersection. In Japan, drivers must turn on their blinkers at least 30 m from thae target intersection (Article 21 of the Order for Enforcement of the Road Traffic Act).

We compare the two methods on a test course, a circuit road in Iwate Prefectural University with intersections made using traffic cones (Figure 6). We first examine an existing navigation system that provides drivers with instructions such as "In 100 m, turn left." A preliminary exploration of a road

Table II. Estimation results for preceived elapsed time

	10 s	20 s	30 s
Average (m)	10.8	21.9	32.5
Avg. difference (m)	1.25	2.6	3.29
Std. dev. (m)	1.44	2.56	3.17

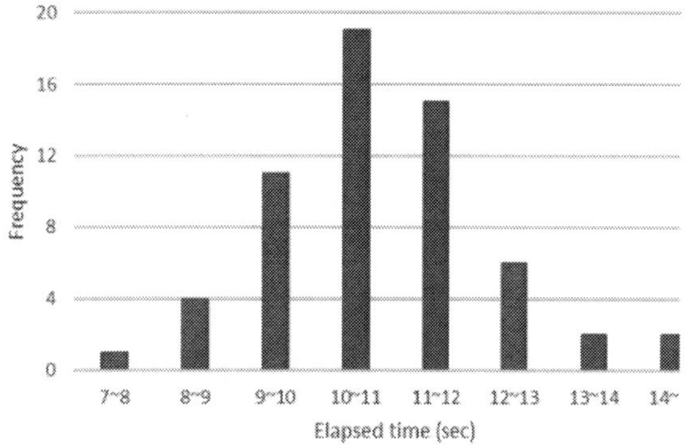

(a) Instruction time = 10 s

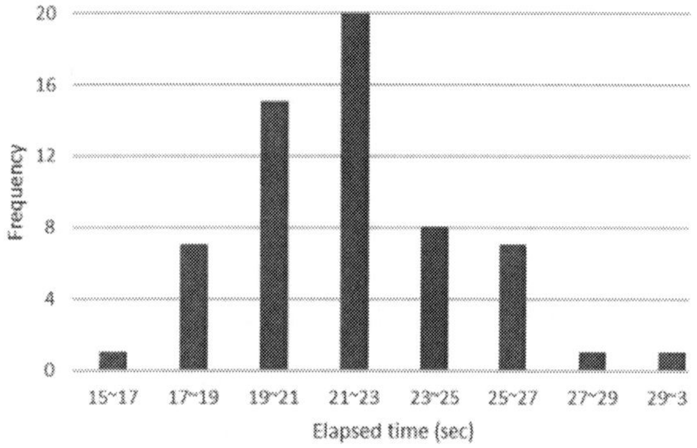

(b) Instruction time = 20 s

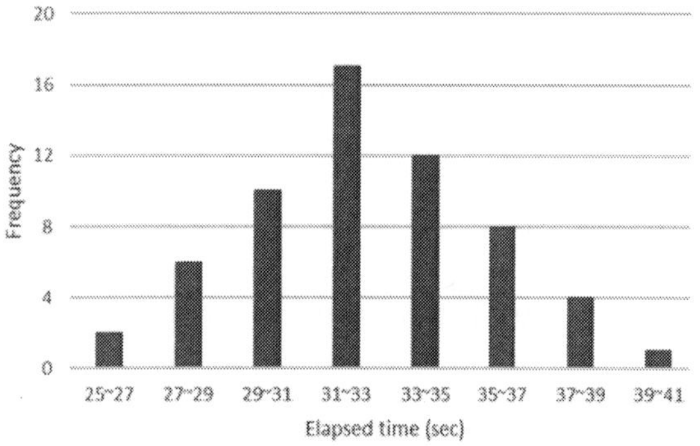

(c) Instruction time = 30 s

Figure 5. Frequency distribution for elapsed time

map suggests that most distances between intersections are less than 20 m and thus, all traffic cones are placed at 20 m intervals. Figure 7 shows an image of the test car (Nissan Micra) and that of the emergency brake installed on the examiner's side for safety purposes. A total of other 12 university students with a driving license participated in the comparative study. We instructed that the participants halted

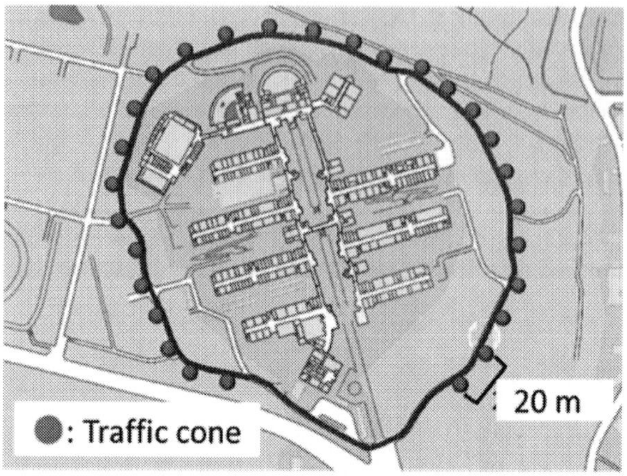

Figure 6. Test course of instruction methods in the university

(1) NISSAN MICRA (2) Emergency brake

Figure 7. Test car

instead of turning when they were notified of an intersection. The test was repeated thrice for each method.

Figure 8 reports the rate of participants who turned (or stopped) correctly in response to each instruction method. The vertical axis denotes the rate of correct turns and the horizontal axis represents each instruction method. The countdown method reports the highest rate of correct turns (91.7%). However, some participants perceived the repeated instructions to be distracting. The rate of correct turns is marginally lower for the blinker method (86.3 %). The participants stated that this method's instructions are simpler than those of the countdown method. As previously mentioned, the blinker method ensures that those in subsequent cars are aware of the driver's intended actions and thus, is superior in term of driving safety.

The blinker method provides the most effective voice instructions, particularly when the distance between intersections is less than 100 m. Figure 9 is an example of a test course for the blinker method. The evaluated distance between the intersections is less than 20 m. A total of 10 participants with a driving license were asked to drive a Nissan Micra.

For the blinker method, the examiner selected three out of six intersections shown in Figure 9. All participants made an equal number of correct right and left turns (i.e., 15 each).

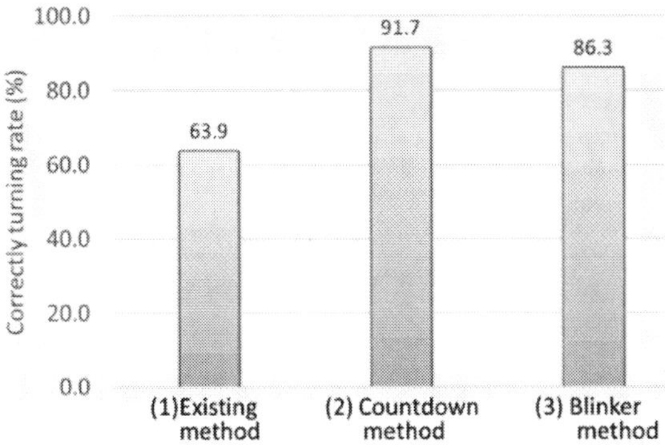

Figure 8. Correctly turning rate

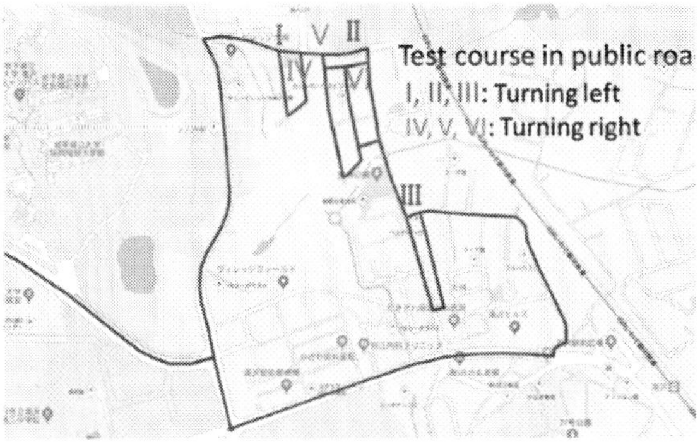

Figure 9. Test course for the blinker method in public roads

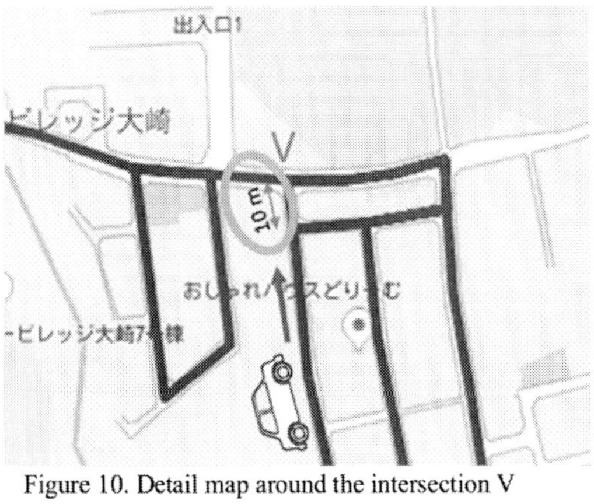

Figure 10. Detail map around the intersection V

However, one participant selected a wrong intersection. This is possibly because the distance between intersections VI and V was only 10 m (see Figure 10), making the intersection difficult to identify given the significantly short distance.

Next, we compare the blinker method with Google Maps on a public road (Figure 9). The test format is similar to that

TABLE III. ACCURACY OF TURNING INTERSECTIONS

Intersection #	Turn L/R	Blinker method (%)	Google Map (%)
I	L	100	100
II	L	100	20
III	L	100	40
IV	R	100	80
V	R	80	40
VI	R	100	100

employed for the blinker method. The same 10 drivers were asked to participate in the evaluation of the blinker method. Table III presents the examination results and shows that the proposed blinker method is superior to Google Maps on roads where the distance between intersections is less than 20 m.

V. CONCLUSIONS

Automotive navigation systems have become increasingly popular and accessible through applications on mobile devices such as smartphones. While they help drivers navigate through unfamiliar regions, drivers may occasionally misjudge or make incorrect turns, particularly in response to voice instructions such as "In XX meters, turn right/left." This miscalculation can be attributed to the difference in perceived and actual distance. Thus, this study experimentally evaluates the difference between perceived and actual values for distance and elapsed time. The data indicate that re-instructing drivers regarding distance is effective on urban roads and when the distance is less than 100 m, and instructions related to elapsed time are useful on regional road such as highways.

We propose a guidance method in which drivers are instructed to turn on their right or left blinker upon approaching a turn. This method not only decreases the likelihood of incorrect turns but also is superior in terms of safety driving because it informs the cars behind of the driver's intended actions.

We employed vibrotactile actuators for notifications and will continue to do so for further analyses. Future research could explore guidance methods without sound and visual notifications.

ACKNOWLEDGMENT

This work is financially supported by JSPS KAKENHI [grant no. 16K00276]. We would like to thank Uni-edit (https://uni-edit.net/) for editing and proofreading this manuscript.

REFERENCES

[1] Google, Google Maps, Available from: https:// https://www.google.co.jp/maps/ 2020.10.13

[2] Laura Dal Bo, Jens Lembke, Jörg Lemme, METHODS AND SYSTEMS FOR PROVIDING AN IMPROVED MANEUVER COUNTDOWN BAR, https://www.freepatentsonline.com/y2020/0096358.html, United States Patent Application 20200096358, Kind Code:A1, 2020.

[3] B. S. Jensen, M. B. Skov, and N. Thiruravichandran, "Studying Driver Attention and Behaviour for Three Configurations of GPS Navigation in Real Traffic Driving," ACM, CHI 2010, Proceedings of the SIGCHI Conference on Human Factors in Computing Systems, pp. 1271-1280, 2010.

[4] O. Wiles, M. Mahmoud, P. Robinson, E. Dias, L. Skrypchuk, "Towards a User-Centric In-Vehicle Navigational System" Proceedings of the 8th International Conference on Automotive User Interfaces and Interactive Vehicular Applications, Oct. 2016, pp. 27-34.

[5] Panasonic, Strada CN- F1XVD Available from: https://panasonic.jp/car/navi/products/F1XVD/navi/navi03.html 2020.09.30 [in Japanese]

[6] K.Akaho, T. Nakagawa, Y. Yamaguchi, K. Kawai, H. Kato and S. Nishida, "A Study and Evaluation on Route Guidance of a Car Navigation System Based on Augmented Reality," International Conference on Human-Computer Interaction; Towards Mobile and Intelligent Interaction Environments, Jul. 2011, pp. 357-366..

[7] S. T. Shahriar and A. L. Kun, "Camera-View Augmented Reality: Overlaying Navigation Instructions on a Real-Time View of the Road," ACM, AutomotiveUI 2018, pp. 146-154, 2018.

[8] Pioneer new CYBER NAVI Car Navigation system AVIC-VH99HUD introduction, https://www.youtube.com/watch?v=FODe8JMniX8, [retrieved: September 2020]

[9] D. R.Large and G. E.Burnett, "The effect of different navigation voices on trust and attention while using in-vehicle navigation systems," ELSEVIER, Journal of Safety Research, Vo. 49, pp. 69-75, 2014.

Vibro-tactile Notification in Different Environments for Motorcyclists

Yuta Yamauchi

Graduate school of Software and
Information Science
Iwate prefectural University
020-0693,Sugo Japan
Email: g231r028@s.iwate-pu.ac.jp

Akimasa Suzuki

Department of Information Science,
Iwate prefectural University
020-0693,Sugo Japan
Email: suzuki_a@iwate-pu.ac.jp

Abstract—**This paper evaluates the effectiveness of vibro-tactile notification for motorcyclists under external factors. Although many car manufacturers provide side and rear collision warning systems with auditory or visual alarms, the notifications may confuse a motorcyclist because they already need to be aware of many visual targets such as mirrors and monitors, and environmental sounds. This paper proposes vibro-tactile notification system using a vibration speaker installed in a motorcycle helmet between the outer shell and the cushion. The proposed system should enable motorcyclists to correctly identify the directions of five vibrating motors, three level of risk, and three obstacle types (i.e., pedestrians, vehicles, and motorcycles). We evaluate the system under windy and engine vibration conditions and examine accuracy of notification via experiment. Our results indicate that motorcyclists can correctly detect four directions and three threat levels using this system.**

Keywords–vibro-tactile notifications; helmet actuators; vibration speakers;

I. Introduction

Because once the motorcyclist is in a crash, they are more likely to die as a result of less protection from the vehicle. For motorcyclists, hazard notification is vital, because of their limited visibility and the diverse sounds they may hear and their very high risk of accident. The fatality rate in crashes for motorcyclists is 1.22% , while that for drivers of four-wheeled vehicles is 0.35% [1] . Furthermore, motorcycles are small and difficult for other drivers to recognize. Motorcyclists therefore need to be highly aware of their surroundings, but this is difficult because of the blind spots due to their helmet and small mirrors. To avoid incidents, an intuitive notification system that can specify direction and threat level is required. Therefore, we propose a system that uses haptic sensations to quickly notify motorcyclist of possible hazards or obstacles around the vehicle.

Our proposed hazard notification system uses vibro-tactile actuators installed in a motorcycle helmet. The system notifications flag the type of object, direction, and threat level surrounding the vehicle. We evaluate robustness against wind and motorcycle engine vibration. We also perform experiments to test the effectiveness of our proposed system. Section II presents related research. Section III presents system architecture considerations. Section IV presents the system architecture. Section V presents an examination of the vibration intensity to inform. Section VI presents the experiments under the influence of motorcycle engine vibration and the results of the experiments. Section VII presents the experiments under the influence of the driving wind and the results are presented.

II. Previous and Related Works

For preventing accidents on motorcycles, there are two main approaches: motorcyclists are assisted in checking their surroundings or drivers around a motorcycle are assisted in recognizing motorcyclist locations. For the latter approach, a helmet with brake lights has been investigated for practical user [2]. However, we focus on the former approach in this study. Hazard notification systems have been proposed for four-wheeled vehicles [3] [4].

In addition, sensor systems, such as collision [5] or ground [6] detection methods, have been investigated for detection around a motorcycle. Many systems focused on the sensor, however, we would not sense issues, but also provide information to the motorcyclist because of lacking of the system. One study [7] proposed a smart helmet using a multimedia Internet of Things (IoT) sensor device and visual notifications. Many conventional notification methods for motorcycles rely on visual images in the motorcyclist's view [8] [9], such as front view, mirror, tachometer, speedometer, navigation system, and indicators. Therefore, there is the potential that excessive visual information may instead impact the motorcyclist's capacity to adhere to safe driving practices.

We focus on vibro-tactile notification as non-visual information. We could know the information by vibro-tactile because ancient motor cycles tells engine failure by irregular vibration to riders. Some systems vibrate a motorcycle's steering, but this approach is limited for notifications in front or behind the motorcycle [10]. We previously proposed a system for four-wheeled vehicles that uses haptic sensations to quickly notify drivers of possible hazards or obstacles surrounding the vehicle [11]. We examined the system's robustness against the different material types and layers used for the driving seat cushions of four-wheeled vehicles [12]. Seat on the motorcycle is vibrated by the engine, and it may be more difficult to notify by the vibration. In this study, we perform experiments under wind and engine vibration conditions to consider the viability of a highly intuitive notification system for motorcycles.

III. Mounting positions of actuators for motorcycles

In this section, we represent the system architecture.

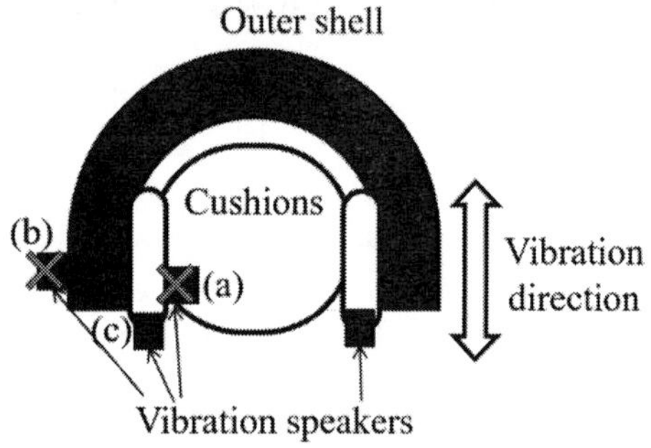

Figure 1: Proposed installation positions for actuators in a motorcycle helmet.

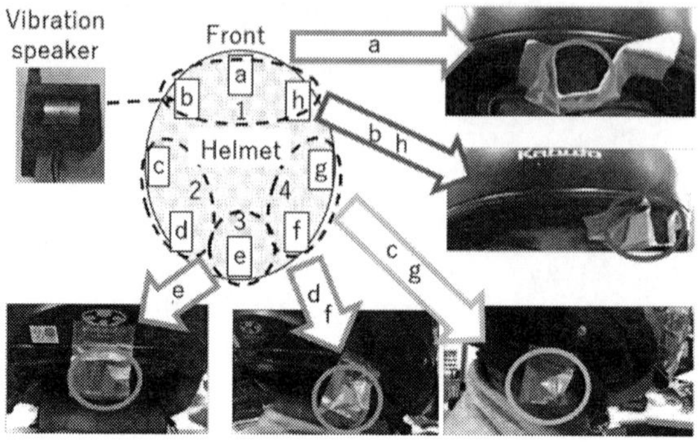

Figure 2: Experimental arrangement of mounted actuators.

A. Notification Parts of a Body by Vibro-Tactile Actuators

The proposed system aims to help motorcyclists correctly identify the directions of five vibrating motors, three intensity settings as the level of risk, and three obstacle types (i.e., pedestrians, vehicles, and motorcycles). The position of vibration is important for giving critical notifications. We can consider vibro-tactile actuators on the motorcyclist's arms, shoulders, or waist within a motorcyclist's suit or on the motorcyclist's head within a helmet. For locations on the body, the motorcyclist must use a wearable device, and there are different types of suit for different motorcycles types (e.g., Cruiser or Sport) because the riding posture is different. Drivers may also experience limited mobility in a suit, which would impact accurate notification.

By contrast, helmets are usually fitted to the motorcyclist's head, and vibration positions are not affected by varying postures, although they are limited to facing forward. Helmets are also required in many countries. Therefore, we choose helmets for our proposed system.

Figure 1 shows the possible installation positions within a motorcycle helmet. We cannot mount an actuator inside the helmet, as at (a), because the direct contact with the driver's head would be unsafe. We considered the helmet surface, as at (b), but our attempts showed that a very strong vibration would be needed. Therefore, we mount the actuators at the bottom of the cushion in the helmet, as at (c), so that the actuators vibrate vertically to each cushion.

We considered three types of vibration mechanisms: vibration motors, haptic reactors, and vibration speakers. Vibration motors can only produce sine waves and it is difficult to distinguish different categories, although they can achieve strong vibration. Although haptic reactors can realize a variety of vibrations, such as clicking, their vibration is too weak to produce notifications. We therefore propose a vibro-tactile notification system using a vibration speaker which can realize the vibration with strong and varied expression. For our proposed system, we utilized a vibrating speaker with an ACOUSTIC HAPTIC^{TM}actuator developed by Foster Electric Company Limited. The acoustic haptic actuator is a type of woofer that comes into direct contact with the driver's helmet.

B. Mounting Positions of Actuators on a Helmet

Motorcycle helmets have shields that can reduce visibility. They may also become foggy with weather and temperature, which further limits the motorcyclist's visibility. Backward of the motorcycles have more blind spot where such as motorcyclist cannot watch by the mirrors. Therefore, we consider the need for both backward and forward alerts.

Figure 2 shows the planned layout of actuators in the helmet. This helmet has four cushions (i.e., front, rear, right and left), as shown in sections (1) to (4) in Figure 2. To explore the vibro-tactile directional sense at the motorcyclist's head, we installed eight actuators as shown in Figures 2(a) to (h). Actuators (a), (b), and (h) were mounted on cushion (1), (c) and (d) were mounted on cushion (2), (e) was mounted on cushion (3), and (f) and (g) were mounted on cushion (4).

We conducted an experiment to evaluate the resolution of vibration on the human head. By determining the resolution of human-perceivable locations, we can determine which directions are identifiable. We performed the experiment with the engine in idle at 1500 ± 300rpm, which is frequently used as the typically speed range, on a Yamaha MT-01 motorcycle equipped with a V-twin cylinder 1670 cc engine. We conducted four trials with each participant. We apply five students between 19 and 22 years old as the participants. We randomly induced vibrations at each position with strengths ranging from -2dB to -12dB and participants estimated the position in the helmet.

Figure 3 shows the correct answer rates for the direction using all eight or only four actuators (one on each cushion: (a), (c), (e), and (g) in Figure 2). The vertical and horizontal axes denote correct answer rates and installation positions, respectively, and the green and orange bars respectively indicate results for all eight or only four actuators. The results demonstrate that there was confusion when multiple actuators were installed on a single cushion, leading to lower accuracy. By contrast, all participants had a 100% correct answer rate when four actuators were used. We therefore decided to utilize only one actuator per cushion.

IV. System Architecture for Our Proposed System

In line with the aforementioned experiment shown in Figures 2 and 3, we propose a vibro-tactile helmet as illustrated in Figure 4. Figure 4(A) shows the positions of the actuators (or vibration speakers) on the helmet. Figure 4(B) shows pictures of the different actuators in place; note that (b) and (d) are in the same position on the left and right sides of the helmet, so only one is shown.

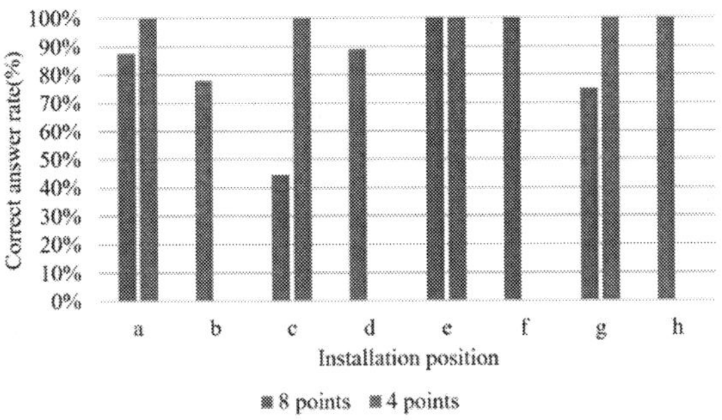

Figure 3: Correct answer rates according to the number and position of actuators.

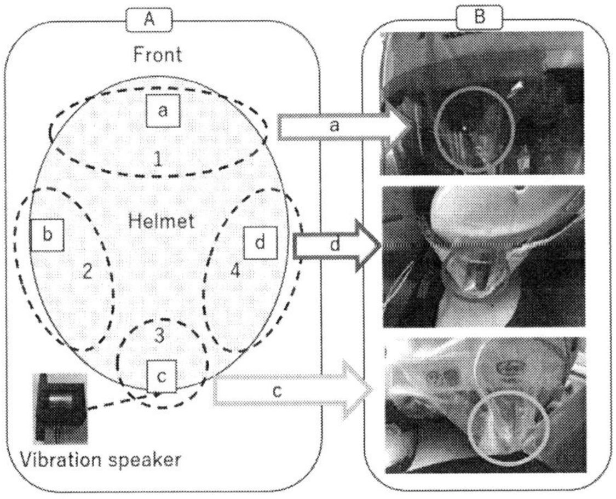

Figure 4: Overview of our proposed system.

The actuators shown in Figure 4 vibrate by transmitting sound data via the amplifiers. The sound data are deformed waves for three categories (i.e., pedestrian, four-wheeled vehicle, and motorcycle) the same as in our previous study for four-wheeled vehicles [12].

V. Vibration Strength

In this section, we discuss the strength of the vibrations for our proposed system.

A. Estimation of Strength Levels

Our previous study in four-wheeled vehicles [11] had three vibration strength levels (i.e., large, medium, and small) as the level of risk. However, when riding a motorcycle, there is additional noise, such as from engine vibration or wind. Furthermore, we have not yet applied the vibration strength level concept to a helmet. Therefore, we consider the vibration strength of the three levels.

An experiment was conducted with five participants between 19 and 22 years old and the Yamaha MT-01 in idle. The strength pattern is defined by the difference in the sound pressure. We considered two strength patterns. Pattern A has a small difference between the three levels. We can utilize a fourth level if we can detect the differences in Pattern A. Pattern B has larger differences than Pattern A, with the

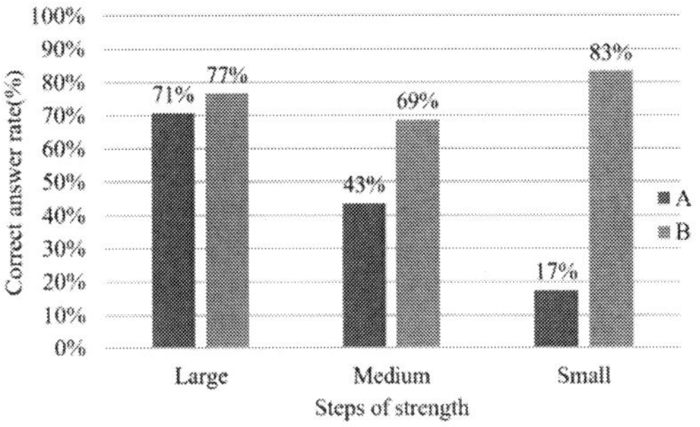

Figure 5: Correct answer rates for strength Patterns A and B.

larger vibration for four-wheeled vehicles adjusted so as not to prevent the motorcyclist from driving.

The strength levels of Pattern A, "large", "medium", and "small", were respectively -6dB, -10dB, and -12dB from the original sound data which used by previous study [8], for all categories. The strength levels of Pattern B were different according to the category. In the cases of the pedestrian or motorcycle categories, "large", "medium", and "small" were respectively -2dB, -8dB, and -12dB from the original sound data. In the case of four-wheeled vehicles, "large", "medium", and "small" were respectively -3dB, -8dB, and -12dB from the original sound data.

This experiment was conducted using random directions, categories, and vibration strengths. Test participants answered "large", "medium", "small", or "insensitive" as the levels of strength. Figure 5 shows the experimental results for both patterns. The vertical and horizontal axes denote correct answer rates and strength levels, respectively. The blue and brown bars respectively indicate Patterns A and B. Pattern A had low correct answer rates for "medium" and "small", and as well as instances of "insensitive" shown as 0 percent in Figure 5. Pattern B had correct answer rates of over 69% for each strength level. Thus, we adopt the notifying method by three strength levels with large interval such as Pattern B.

B. Normalization by Head Sense

In our previous study, we improved correct answer rates by using normalized and exaggerated waves for vibrations for the three categories (i.e., four-wheeled vehicles, motorcycles, and pedestrians) [12]. In this study, we similarly apply normalized and exaggerated waves to improve notification accuracy because many motorcyclists pointed out that vibration strength felt uneven depending on the installation position. Here, we normalize the vibration strength for the parts of the head an experiment on the motorcycle.

For normalization, we utilized three actuators, (a), (b), and (c), on the front, rear, and left cushions in Figure 4, respectively. Actuator (d) in Figure 4 is considered to have the same tendency as actuator (b). First, via a questionnaire, we determined the maximum and minimum strengths motorcyclists can detect with no stress. The results indicate that the difference between the maximum and minimum strengths for actuator (b) was smaller than that for the other positions. Thus, we used the maximum and minimum strengths, which participants feels as same as (b), for all the actuators. The maximum and minimum strengths were defined as "large"

TABLE I: VIBRATION STRENGTH LEVELS AT EACH POSITION.

position	large	medium	small
a	−6dB	−10dB	−14dB
b	−8dB	−12dB	−19dB
c	−6dB	−10dB	−13dB

and "small". The "medium" strength level was not defined as the midpoint value (in decibels) between "large" and "small", but defined as participants feel "medium" between "large" and "small" vibration. In this normalization, we utilized the vibration of the four-wheeled vehicle category [12]. Finally, we adjusted for the strength as shown in Table I, which presents the strength levels for each actuator position, noting that the side position is more sensitive than the front and back positions.

VI. Experiment

Finally, we conduct an experiment to verify the correct answer rates when using actuators mounted on a helmet when there are external factors, including wind and engine noise.

A. Experimental Trials by t-test

We decided to use t-tests to determine statistically significant results. We conducted an independent t-test for each strength level on actuators (a), (b), and (c) to compare the differences in correct answer rates under different wind and idling noises. We adopted a significance level of 5%, a moderate effect size of 0.5, and a detection rate of 80%. Possible answers were "large", "medium", "small", and "insensitive". Sample sizes for the answer of "large" and the other strength levels were determined from one-sided and two-sided tests, respectively. From the t-test, sample sizes from which we could obtain a significant difference were 51, 64, and 64 samples for "large", "medium", and "small" answers, respectively.

B. Correct Answer Rates during Idling

We conducted an experiment in the idling state to evaluate the robustness under engine vibration, as illustrated in Figure 6. Six test participants on the motorcycle answered when they felt a vibration. This experiment was conducted with the engine off as 0rpm, and rotating at 1000rpm, 1500rpm, and 2000rpm. In the case of 0rpm/h, the experiment was considered as a stable situation. For each engine speed include 0rpm/h, 51, 64, and 64 trials for "large", "medium", and "small", respectively, were performed at random. The vibration categories used were those applied to four-wheeled vehicles in our previous study [12].

Figure 7 shows the correct answer rates during idling rpm at 0rpm, 1000rpm, 1500rpm, and 2000rpm, respectively. The blue, red, and yellow bars respectively indicate "large", "medium", and "small" as answered by the participants. The vertical and horizontal axes of each figure denote the correct answer rate and the three levels of signal strength, respectively. For example, 10% of participants answered "medium" for the "large" strength level given by an actuator at 1000rpm, as shown in Figure 7(b).

We defined the correct answer rate as the percentage of matches between the answers of participants and the actual level of vibration strength. For example, in the case of Figure 7(b), the correct answer rates for "large", "medium", and "small" were 90%, 72%, and 83%, respectively. Let us focus on the "medium" strength level in Figure 7. We can confirm that participants felt as more strong vibration identifying the "medium" strength level because participants answered "large" more often than they did "small". The correct answer rates were lowest in the case of Figure 7(d). This may be due to the high engine rotation causing stronger vibration and noise from the motorcycle, obscuring the vibration from the actuators.

Figure 6: Overview of the experiment for the stable situation.

VII. Evaluation of Driving Scenario

We conducted an experiment with six participants using a car to evaluate degradation in accuracy due to wind. We used the car because of safety and difficulty to collect correctly answers of notification. Although the strength of the traveling wind is little different between cars and motorcycles, the effect of the wind can be measured. We compared the wind noise between the car and the motorcycle, and found that the wind noise on the motorcycle was almost same as on the car with all windows open. Each participant evaluated the four speeds of 0km/h, 60km/h, 80km/h, and 100km/h. At each speed, we performed 51, 64, and 64 trials with "large", "medium", and "small" strength levels, respectively, at random. The actuator vibrated for 5–10 s at random for each trial. In the case of 0km/h, we conducted the experiment as a stable situation, as same as section 6.2. In the other cases, we used a highway. Figure 8 shows the experimental highway route. This route has two lanes in each direction limited to 100km/h, and three entrances and exits shown as (1), (2), and (3). We set up sections of 11.1km between (1) and (2), 20.3km between (2) and (3), and 9.2km between (3) and (1).

The experiment was conducted at speeds of 80km/h, 100km/h, and 60km/h between (1) and (2), (2) and (3), and (3) and (1), respectively. Figure 9 shows the seating positions of the participant in this experiment. All windows of the car were open and the helmet shields were closed. Before the experiment, participants were provided with examples of the three strength levels (i.e., "large", "medium", and "small") at position (b) of 4. We limited the experiment to the four-wheeled vehicle category. The experimental results were saved as movie files and evaluated via post-processing.

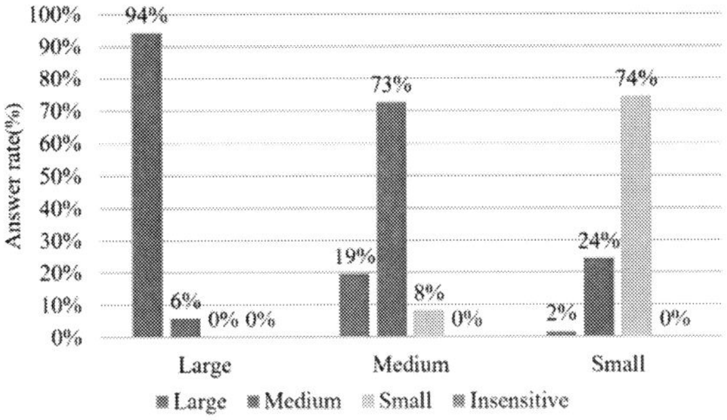

(a) 0rpm

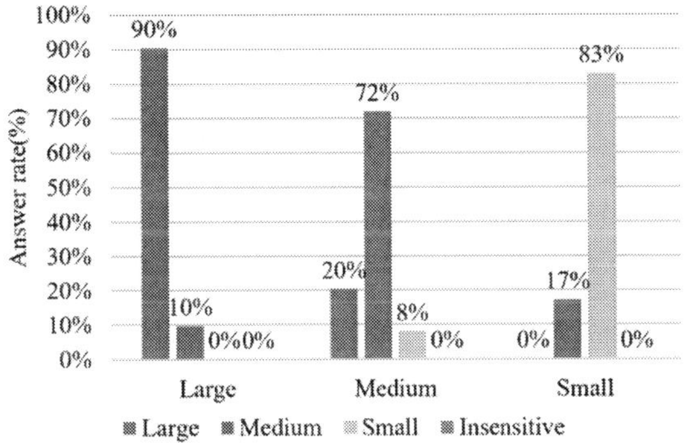

(b) 1000rpm

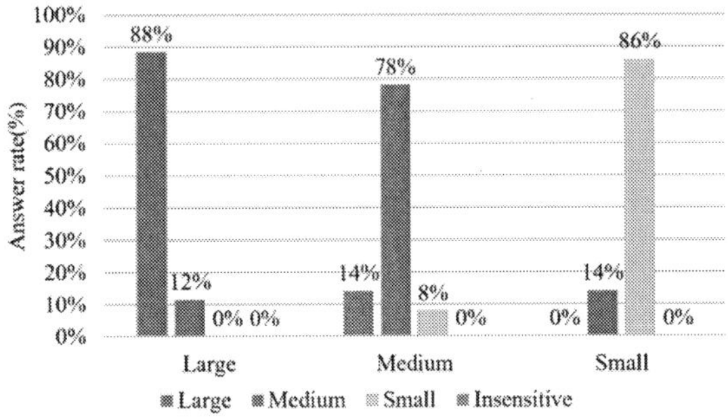

(c) 1500rpm

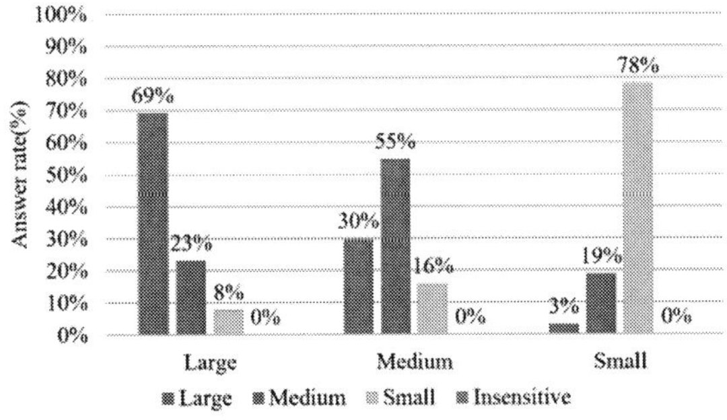

(d) 2000rpm

Figure 7: Answer rates for different strength levels at (a) 0 rpm, (b) 1000 rpm, (c) 1500 rpm, and (d) 2000 rpm.

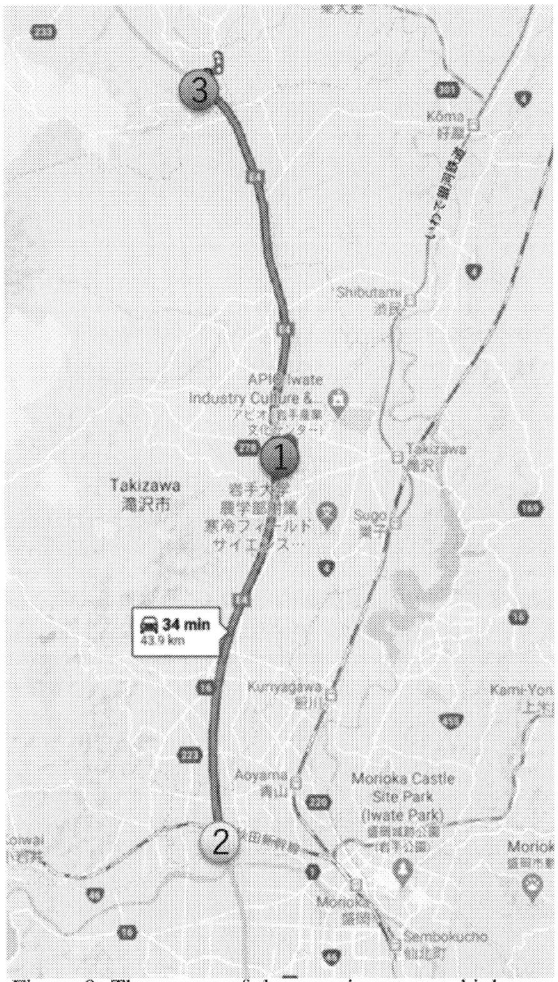

Figure 8: The course of the experiment on a highway.

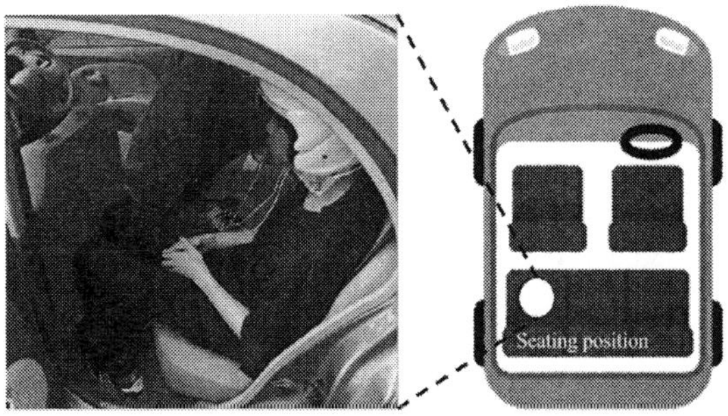

Figure 9: The seating position of participants in the vehicle.

Figure 10 shows the correct answer rates at driving speeds of 60km/h, 80km/h, and 100km/h. The vertical and horizontal axes of each figure denote the correct answer rate and the three levels of signal strength, respectively. In the case of Figure 10(a), in 94%, 73%, and 74% of all trials, the test participant answered "large", "medium", and "small" for large, medium, and small vibration strengths, respectively, so these are the correct answer rates. Figure 10(b) indicates a high correct answer rate for large vibration. Test participants tended to be more likely to select strong vibrations. In the case of Figure 10(c), the correct answer rates were increased and decreased, respectively, for medium and large vibrations as compared to Figure 10(b). In Figure 10(d), a high correct answer rate was obtained even for small vibration. From Figures 10(c) and 10(d), the "medium" strength level showed only small differences in the incorrect answer rate as compared to "large" and "small". Therefore, "medium" is considered to be appropriate in the high-speed case. We expect a higher notification accuracy can be achieved by adjusting the strength automatically depending on outside noise. We also found that notification accuracy was more degraded by engine rotation than by wind noise, which should be a consideration for practical implementation.

VIII. Conclusions

It is difficult for motorcyclists to recognize objects in their surroundings because of the many blind spots from their helmets and small mirrors. Furthermore, accidents are more serious for motorcycles than for four-wheeled vehicles because of the mortality. We therefore proposed a notification system for motorcycles based on previous works for four-wheeled vehicles. In our system, parts of the helmet vibrate corresponding to direction of a hazard, the category of an object, and the level of risk. We considered the strength of vibration to determine three strength levels. We evaluated the accuracy of our proposed notification method for motorcycles using haptic actuators in windy and idling situations. We demonstrated the effectiveness of our notification method even for winds of 100km/h. We expect improved notification accuracy can be achieved by adjusting vibration strength according to the motorcycle's speed. Various types of helmets will be studied in the future.

Acknowledgements

This work was supported by MEXT KAKENHI Grant Number JP 16723884 and Foster Electric Company, Limited. We would like to thank Uni-edit (https://uni-edit.net/) for editing and proofreading this manuscript.

References

[1] "Statistics about road traffic annual report, national public safety commision and national police agency," 2018. [Online]. Available: https://www.e-stat.go.jp/en/stat-search/files?page=1&layout=datalist&toukei=00130002&tstat=000001027457&cycle=7&year=20180&month=0&stat_infid=000031800894

[2] "Brake free helmet light. digital image. indiegogo," 2020. [Online]. Available: https://www.indiegogo.com/projects/brakefree-the-smart-brake-light-for-motorcyclists#/

[3] W. Chang, W. Hwang, and Y. G. Ji, "Haptic seat interfaces for driver information and warning systems," International Journal of Human-Computer Interaction, vol. 27, no. 12, Dec. 2011, pp. 1119–1132. [Online]. Available: https://doi.org/10.1080/10447318.2011.555321

[4] Y. Gaffary and A. Lécuyer, "The use of haptic and tactile information in the car to improve driving safety: A review of current technologies," Frontiers in ICT, vol. 5, Mar. 2018, p. 5. [Online]. Available: https://doi.org/10.3389/fict.2018.00005

[5] M. Muzammel, M. Z. Yusoff, and F. Meriaudeau, "Rear-end vision-based collision detection system for motorcyclists," Journal of Electronic Imaging, vol. 26, no. 3, May 2017, p. 033002. [Online]. Available: https://doi.org/10.1117/1.jei.26.3.033002

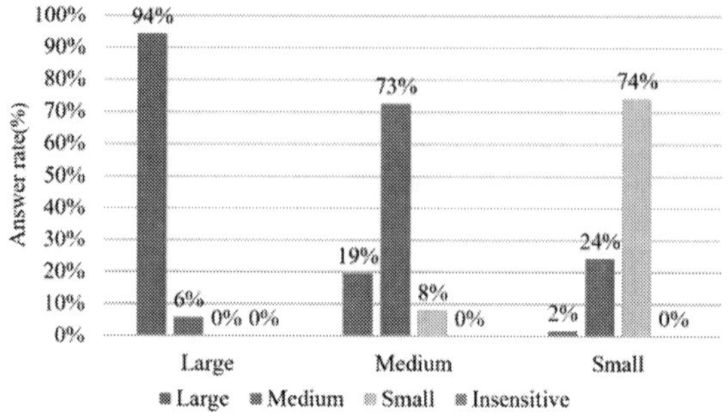

(a) 0km/h

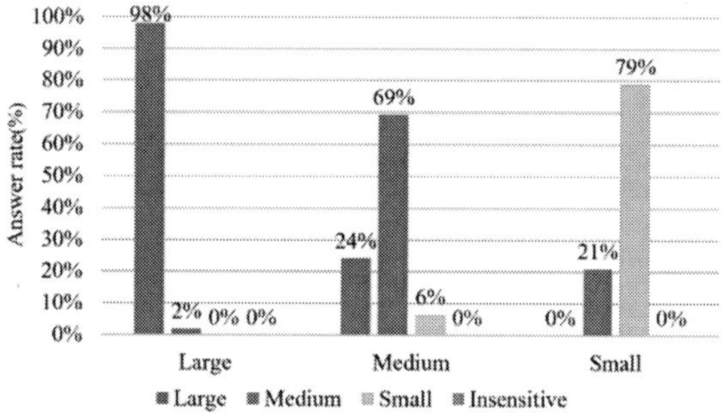

(b) 60km/h

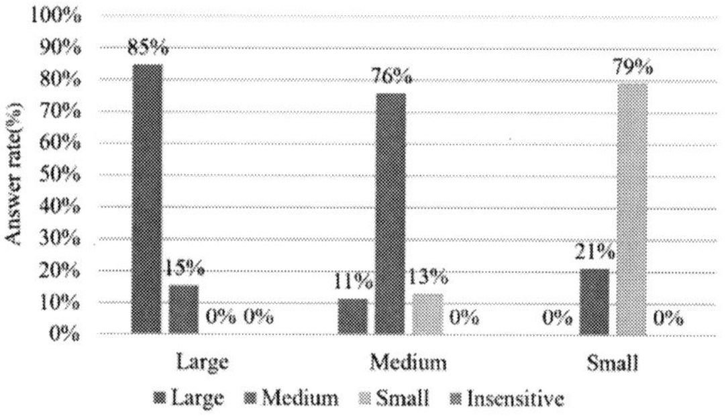

(c) 80km/h

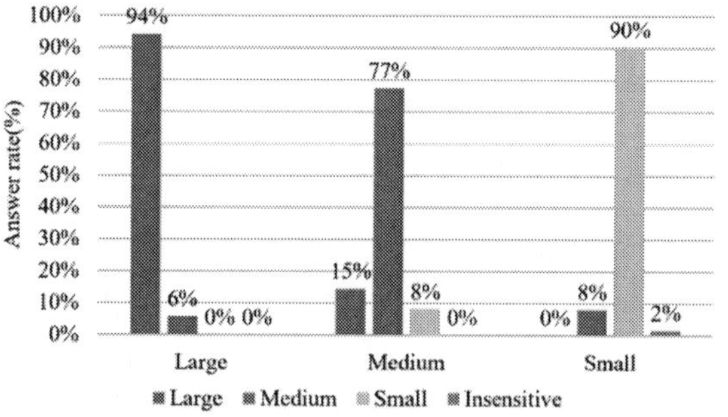

(d) 100km/h

Figure 10: Answer rates for strength levels at each speed.

[6] G. Gil, G. Savino, S. Piantini, and M. Pierini, "Motorcycles that see: Multifocal stereo vision sensor for advanced safety systems in tilting vehicles," Sensors, vol. 18, no. 2, Jan. 2018, p. 295. [Online]. Available: https://doi.org/10.3390/s18010295

[7] M. I. Tayag and M. E. A. D. V. Capuno, "Smart motorcycle helmet: Real-time crash detection with emergency notification, tracker and anti-theft system using internet-of-things cloud based technology," International Journal of Computer Science and Information Technology, vol. 11, no. 03, Jun. 2019, pp. 81–94. [Online]. Available: https://doi.org/10.5121/ijcsit.2019.11307

[8] V. H. Tran, D. Muscat, J. R. Cruz, A. M. Carrillo, D. A. A. Alonso, and E. Chavez, "Motorcycle and helmet providing advance driver assistance," U.S. Patent Application No. 13/897,570, 2013.

[9] Crosshelmet, "Cross helmet x1 - transform your riding experience," 2020. [Online]. Available: https://crosshelmet.com/

[10] C. Inc., "Blind spot detection & lane change assist," 2020. [Online]. Available: https://www.continental-automotive.com/en-gl/2-Wheeler/Safe-Mobility/ARAS/Blind-Spot-Detection

[11] A. Suzuki, Y. Murata, and M. Hayashi, "Notification of hazards around a vehicle using seat actuators," in 25th ITS World Congress. Copenhagen, Denmark: ERTICO, September 2018, pp. AP–TP1335.

[12] A. Suzuki, K. Horie, S. Otobe, Y. Murata, and S. Fujimura, "Hazard notifications around a vehicle using seat actuators," The International Journal on Advances in Intelligent Systems, vol. 13, no. 1 and 2, 2020, pp. 59–68, iSBN:978-1-61208-720-7,ISSN:1942-2679.

Measurement Accuracy on Indoor Positioning System Using SS Ultrasonic Waves for Drone Applications

Tatsuki Okada

Graduate school of Software and
Information Science
Iwate Prefectural University
020-0693,Sugo Japan
Email: g231s004@s.iwate-pu.ac.jp

Akimasa Suzuki

Department of Information Science,
Iwate Prefectural University
020-0693,Sugo Japan
Email: suzuki_a@iwate-pu.ac.jp

Abstract—**This study develops a drone positioning system for use in indoor environments, including in dark places, inaccessible areas, and ordinary living environments where it is difficult to realize by any conventional methods. Various indoor applications using drones have been developed for applications such as drone communication systems and wall surface inspection, which require remote estimation of their position. For outdoor applications, a Global Navigation Satellite System (GNSS) is generally used to obtain the drone position. However, as the radiowaves of the GNSS cannot reach indoors or between buildings, camera-based methods, such as Simultaneous Localization And Sapping (SLAM), are applied to estimate the drone's position. The system uses a noise-resistant, code-division-multiplexed Spread Spectrum (SS) ultrasonic waves for three-dimensional positioning. We develop transmitter and receiver hardware using SS ultrasonic waves and evaluate the effect of wind and sound of the positioning system during drone operations on the SS ultrasonic positioning. The accuracy of the positioning system was verified through experiments, and the results showed that a positioning accuracy within 15 cm was possible despite the effects of downwash generated by the drone's wings.**

Keywords–Drone; Indoor Positioning System; SS Ultrasonic Waves; Downwash

I. INTRODUCTION

As Unmanned Aerial Vehicles (UAV), drones can be flown autonomously or operated by remote control. Because they can take off and land vertically in small spaces, they can be used to perform a variety of activities in unstable places where people and vehicles cannot enter. Previous studies have investigated the use of drones for autonomous search and rescue operations for victims following a disaster [1], meteorological observations [2], and logistics such as home delivery [3].

When used indoors, drones act as communication robots [4]. However, an appropriate distance is required to allow natural and smooth communication between a human and an autonomous mobile robot. To ensure the appropriate positioning in indoor spaces, the drone's coordinates can be used to develop real-time centimeter-order positioning. On the other hand, a relevant study investigated the use of drones for periodic inspections to detect aging degradation of locations where staff is unable to work, such as high walls of tanks and industrial chimneys [5]. Using drones to conduct periodic inspections is expected to reduce the high cost of these inspections.

It is more dangerous to use drones indoors than outdoors because it is easier to crash the drone into obstacles, such as humans and walls. Thus, it is essential to determine the position of the drone in relation to other objects. As horizontal and vertical relationships are important in these applications, it is essential to obtain absolute coordinates in space. While a Global Navigation Satellite System (GNSS) is generally used to obtain the absolute coordinates of a drone, the GNSS signal is difficult to detect indoors. Simultaneous Localization and Mapping (SLAM) is often used in non-GNSS environments. However, the flight path of a routine inspection is often in a dark place and the walls do not always follow a uniform pattern, causing large errors in SLAM's self-position estimation.

We therefore propose an indoor positioning system for drones using spread spectrum (SS) ultrasonic waves [6]. This system is expected to obtain 3D coordinates with an accuracy of 10cm. However, noise from the propellers or downwash of a drone may lower this accuracy. Downwash is the wind created by the drone's propellers. Therefore, this study conducts an experiment to evaluate the positioning accuracy of drone flights during a periodic inspection. Section II presents related research. Section III presents an overview of indoor positioning systems using SS ultrasound. Section IV presents the experiments and their results.

II. RELEVANT STUDIES AND PREVIOUS WORKS

There is no positioning method with drones for indoor multi environments, including dark environments with accuracy under 10 cm. Various sensor systems have been investigated for indoor positioning purposes, including pseudolites [7] and beacons [8]. Of these, ultrasonic-wave-based systems have lower cost and greater accuracy. However, because these systems use the time-division multiplexing method with on-off keying, which grows increasingly cumbersome as the number of objects to be measured grows, they generally have weak noise resistance and are slow to acquire data. Systems using SS ultrasonic signals have therefore been investigated to overcome these drawbacks [9] [10].

Analogous to SS radiowave systems (e.g., GPS), we have proposed a real-time 3D positioning system using SS ultrasonic signals with a band-limited transducer, A Low-Power Field Programmable Gate Array (FPGA), and a small microprocessor [11] [12]. In previous studies, we discussed factors such as positioning errors in indoor environments [6] and signal degradation with band-limited transducers [13] and showed the measurement accuracy of the positioning system using SS ultrasonic signals. We also proposed a calculation algorithm based on the Newton–Raphson method for continuous signals, rather than conventional pulse signals. As a result, 3D coordinates can be obtained every 80ms using Code Division Multiple Access (CDMA) with continuous signals [14].

We evaluated the positioning accuracy of SS ultrasonic waves using a ground-based mobile robot [15]. Other studies have proposed using not only SS ultrasonic waves but also image sensors for drone positioning [16] and applying drones to limited situations such as a greenhouse [17]. Indoor positioning accuracy has been discussed using the Kinect camera, the average positioning error was 48mm [18]. However, it is difficult to use in the dark. This study develops an indoor positioning system using only SS ultrasonic waves that can be used in dark places where image sensors are ineffective.

III. Indoor Drone Positioning System Using SS Ultrasonic Signals

This section describes the indoor positioning method using SS ultrasonic waves and our proposed system.

A. A method for positional calculation

Figure 1 presents the positioning calculations for the indoor positioning system using SS ultrasonic waves. Spheres are drawn to determine the center point on the radius between a receiver Rc and each transmitter, and two pairs of spheres are selected centering on Tr_1 and Tr_3, and Tr_2 and Tr_3, respectively. From these pairs of spheres, $Plane_{13}$ and $Plane_{23}$ are solved as a simultaneous equation and a line of intersection is obtained from the two planes. Last, the points at the intersection of the line with an equation of an arbitrary sphere are solved. Figure 2 also shows a flowchart of the algorithm for the positioning calculation in Figure 1. Two intersection points are obtained transmitters are installed along the wall; therefore, one solution becomes outside of the room. Thus, the other solution becomes the position of the receiver Rc. When using four transmitters, four position results are obtained. Thus, the measurement position is defined as an average of these results.

B. Hardware structure of a positioning system using SS ultrasonic waves

A 3D position can be calculated on the basis of three or more Times Of Flight (TOF) between the transmitters and the receiver. Figure 3 shows the system architecture of the TOF measurement for the positioning system. The transmission unit contains a D/A converter and an FPGA to generate carrier waves and M-sequences. The reception unit includes an A/D converter and an FPGA for correlation calculation, peak detection, and time measurement.

An SS signal is generated by the transmission unit to multiple carrier waves by M-sequences and is outputted from a transducer after D/A conversions. At the start of the transmission, a time counter is started to measure the TOF, and correlation values are calculated from the sound data via the A/D converter as online and real-time hardware processing. The time counter measures the TOF by counting the sampling times until arriving at peak correlation values, obtained by the peak detector. Then, the 3D position of the receiver can be calculated based on three or more TOFs between the

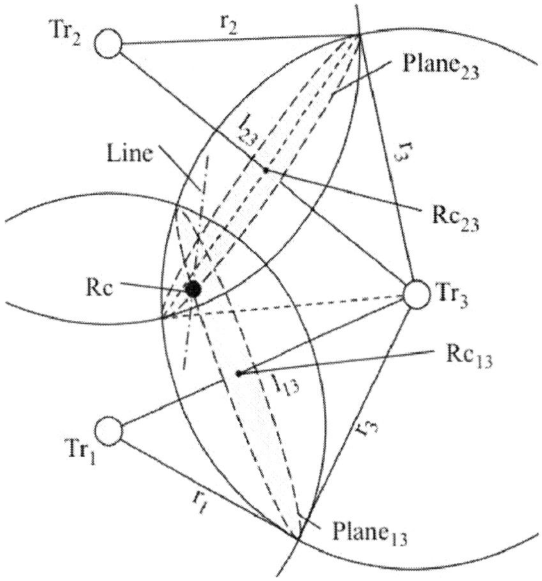

Figure 1. Positioning calculations for the indoor positioning system.

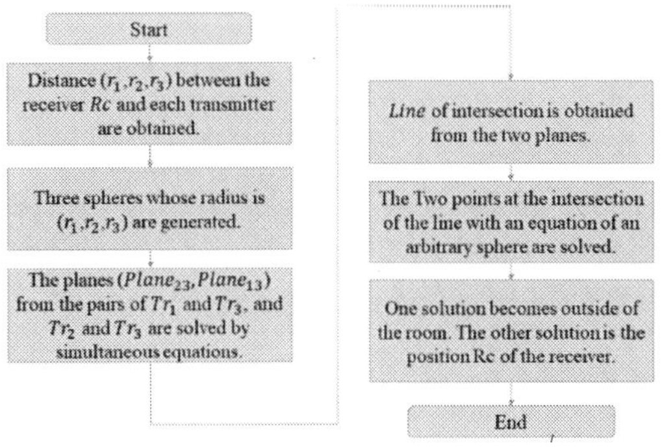

Figure 2. Flowchart on our position calculation

transmitters and receiver. The correlation calculator part is installed in the hardware as shown in Figure 3. Distance is obtained from the TOF obtained from the hardware and dimensional position is measured. Real-time positioning is enough available because this processing can be calculated lightly in software using optimized expressions.

C. SS signal

In our indoor positioning system, SS signals are modulated by binary phase shift keying using M-sequence, a pseudo-random code sequence, with a direct sequence method. Although the M-sequence of '0' or '1' is generated by a shift register, we replace a value of '-1' with '0' for easy signal processing. Figure 4 shows a received SS signal, where the signals corresponding to '1' and '-1' are plotted in solid and dashed lines, respectively. Each dot of Figure 4 is described as a sample convert to digital signal; the amount of sample including one period of carrier waves is decided on four samples. Here, chip length t_c is defined as the time required to describe 1-chip of the M-sequence; the chip length can also be described as $t_c = 4/f$ using carrier frequency f. The length of

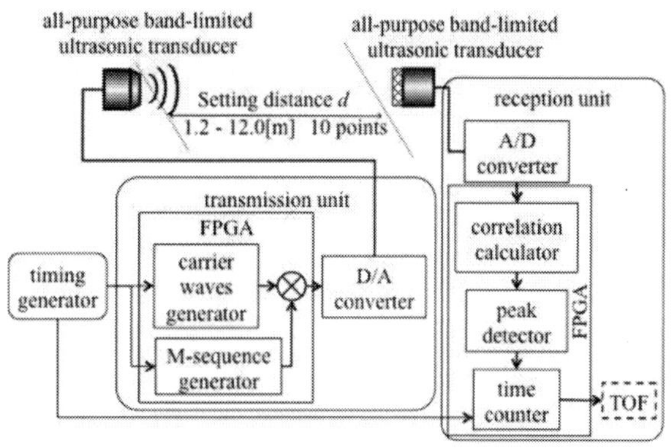

Figure 3. System architecture of the TOF measurement.

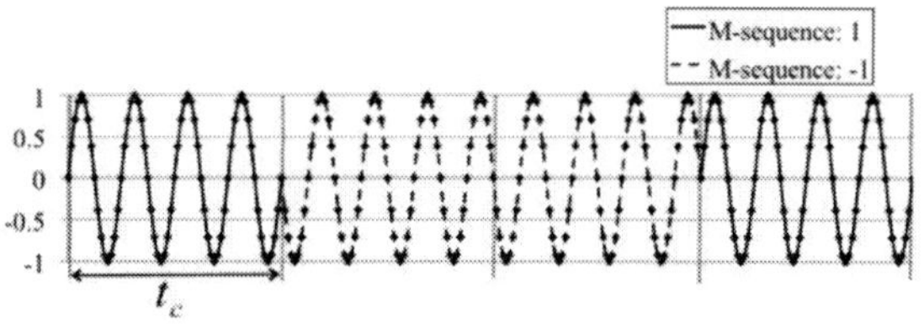

Figure 4. Spread spectrum ultrasonic signal.

SS ultrasonic signals becomes $2^9 - 1 = 511$ [chip] due to a 9-bit shift register for the M-sequence in our system. These four channels of the transmitters are generated by the following tap positions: {4,9}, {3,4,6,9}, {4,5,8,9}, and {1,4,8,9}. In this system, the frequency of the carrier waves is 40.2kHz.

D. Our proposed indoor positioning system using SS ultrasonic for drones

In this system, we use a transmitter with a closed-type aperture (PC40-18S, Nippon Ceramic Co., Ltd.) and a "Mini" SiSonic™ultrasonic receiver (SPM0404UD5, Knowles) as general-purpose ultrasonic transducers.

Figure 5 shows the layout of the transmitters and receiver for our proposed system. We use two example situations, a dark plant and a room, as shown in Figures 5(a) and 5(b), respectively. Figure 5(a) represents a periodical inspection at a plant, where it is difficult to install infrastructure, such as transmitters, in the building. Transmitters are therefore set on a cross-shaped mount, as shown in Figure 5(a), for convenient mounting. Considering the Dilution Of Precision (DOP), the larger the mount size, the more accurate the expected positioning accuracy, although a larger size limits the installation position options and is inconvenient to carry. Figure 5(b) represents a communication drone. Transmitters are mounted in four corners of a room. In this situation, the transmitters are more difficult to install, but the DOP is better than the situation in Figure 5(a). In this paper, we conduct experiments using the layout shown in Figure 5(a).

The drone is fitted with a microphone and small receiving hardware, which processes the ultrasonic waves. Figure 6 illustrates the hardware schematic mounted on the drone. The hardware consists of an Evaluation Board (ACM204-1158C) installed in the FPGA (Intel Cyclone IV); a transceiver for timing the synchronization of the ultrasonic transmitter unit; a receiver unit; a USB interface for output coordinates; an input part to receive the ultrasonic waves, including an A/D converter

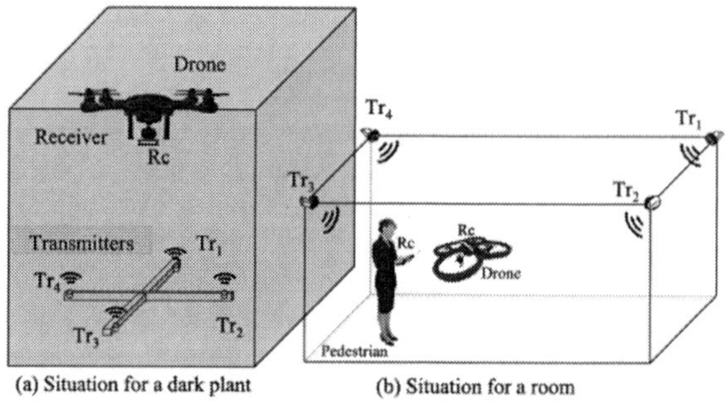

Figure 5. Measurement layout for the proposed system for (a) a dark plant and (b) a room.

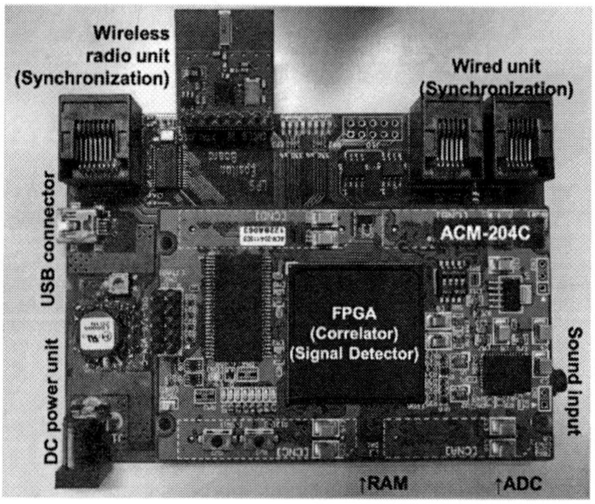

Figure 6. Receiving hardware for the drone measurements.

and amplifier; and SD-RAM for the real-time correlation calculations.

The USB interface and input part are connected to a computer and a microphone, respectively. Ultrasonic waves received by the microphone are converted to A/D and input into the FPGA, where the correlation calculations, peak detection, and TOF calculations are performed. The SD-RAM processes the real-time correlation calculations, and the transceiver measures the TOF based on the transmission timing received from the ultrasonic transmitter. Finally, the drone obtains the TOF by USB UART/FIFO IC (FTDI FT232H).

IV. Positioning Error by Drone Noise and Downwash

We conducted experiments to evaluate the effect of motor noise, wind noise, and downwash generated by the propellers during flight on the SS ultrasonic positioning. Figure 7 shows the environment used for this experiment, which was a room 2000mm long and 4000mm wide. We used a Mavic 2 zoom by DJI™drone for this experiment. Four transmitters Tr_1–Tr_4 were placed near the center of the room. As shown in Figure 7, the drone's starting point was the floor at the left front edge of the room. The coordinates of the transmitters were $\mathrm{Tr}_1[\mathrm{mm}] = (500, 2000, 1500)$, $\mathrm{Tr}_2[\mathrm{mm}] = (1000, 1500, 1500)$, $\mathrm{Tr}_3[\mathrm{mm}] = (1500, 2000, 1500)$, and $\mathrm{Tr}_4[\mathrm{mm}] = (1000, 2500, 1500)$. The transmitting SS signal was amplified to $50\mathrm{V}_{\mathrm{p-p}}$.

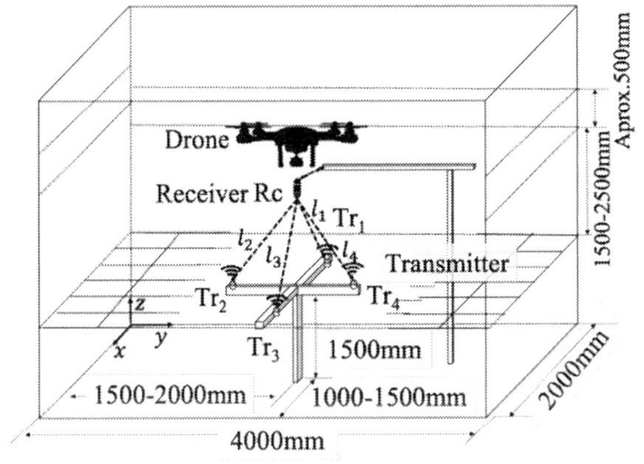

Figure 7. Layout for the positioning experiment.

Figure 8 shows the experimental environment. SS Ultrasonic waves are transmitted upward from Tr_1–Tr_4 mounted on a tripod, and received by Rc, mounted on a bridge of wood.

Figure 9 shows the measurement point. The white and black circles in Figure 9 denote the transmit and receive points, respectively. $\mathrm{Rc}_{(1000,2000,3000)}[\mathrm{mm}]$, $\mathrm{Rc}_{(1000,2000,3500)}[\mathrm{mm}]$, and $\mathrm{Rc}_{(1000,2000,4000)}[\mathrm{mm}]$ are the receiving points at the center of the $x-y$ plane of the four transmitters, located 1500mm, 2000mm, and 2500mm above the transmitters, respectively. Of the other receiving points, $\mathrm{Rc}_{(1000,1500,3000)}[\mathrm{mm}]$ and $\mathrm{Rc}_{(1000,1500,4000)}[\mathrm{mm}]$ are above transmitter Tr_2 and $\mathrm{Rc}_{(500,2000,4000)}[\mathrm{mm}]$ is above transmitter Tr_1.

l_1, l_2, l_3, and l_4, which are the distances between the transmitters and a receiver (Figure 7), are measured for each receiving point. The drone is made to hover at a position approximately 500mm above the receiver. We examine the accuracy when the drone is in flight and when the drone is not in flight in the environment. Five trials were conducted for each receiving point.

A. Measurement error in distance

Figure 10 shows the average differences in the distances from the hovering drone for five trials. The vertical and horizontal axes on each graph denote the difference in distance from the drone compared to the measured distance from l_1 to l_4 to Tr_1 to Tr_4, respectively. The differences in distances are shown as absolute values, and the average difference in the distance is shown as a black line.

The results of the experiment show that all measured distances are obtained when the drone is flying, but the measurement distance is affected by the drone's flight. Figure 10(a) shows the drone's distance for the four transmitters, where the receiving point is on the center of the $x-y$ plane. A greater distance between the transmitter and receiver indicates larger measurement distance. Figure 10(b) compares the accuracy of the distance measurement at the center position $(1000, 2000)$ with that when the drone is above Tr_2 at heights of 3000mm and 4000mm. The difference in distance measurement above Tr_2 is the same as that shown in Figure 10(a). Especially the difference in the distance between Tr_2 and $\mathrm{Rc}_{(1000,1500,4000)}$, where above Tr_2, is increased by the drone hovering. Figure 10(c) shows the measurement distance at the height of 4000mm, which shows that the difference in the distance between Tr_1 and $\mathrm{Rc}_{(500,2000,4000)}$ increases. Compared to $\mathrm{Rc}_{(1000,2000,4000)}$, however, the average difference decreases.

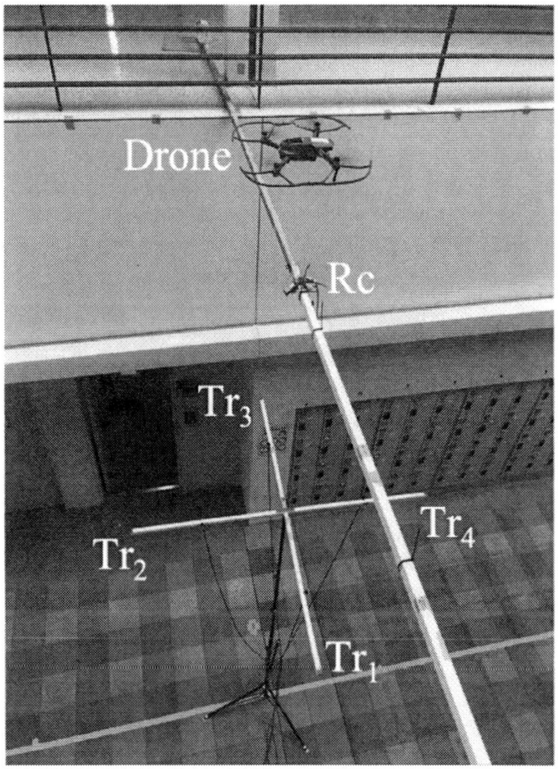

Figure 8. A view of the experiment from above.

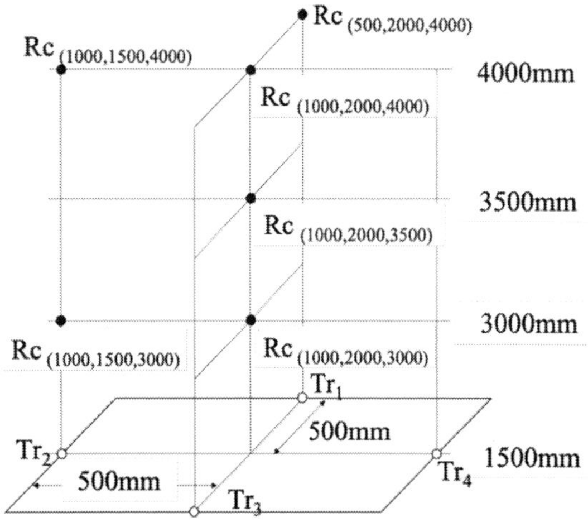

Figure 9. Measurement points of the receivers.

These graphs indicate that a drone's downwash and noise have a significant effect on the measurement distance when the transmitter and receiver are facing each other. The difference in the measured distance with and without drones is within 5cm.

B. Positioning Error

The experimental results were evaluated using the Root Mean Square (RMS) of the difference between the results and

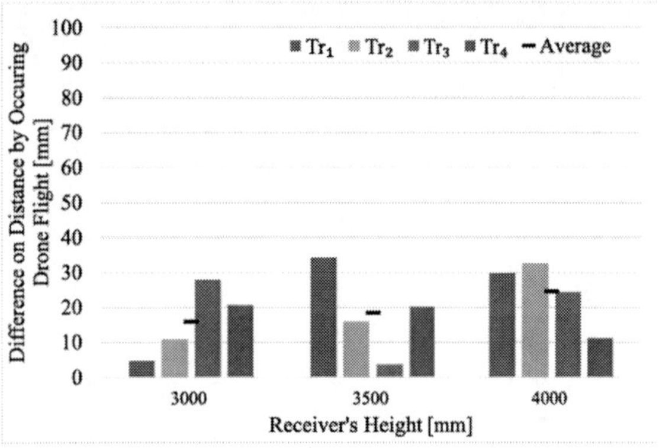

(a) at each receiver's height on the center position

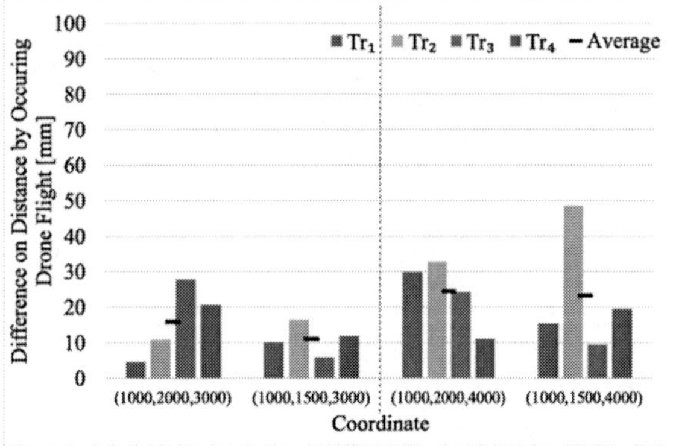

(b) center position vs.above Tr_2

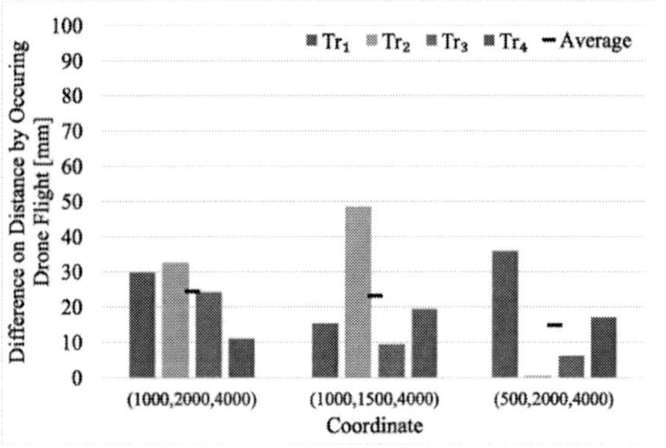

(c) at height of 4000mm

Figure 10. Difference in the measured distance occurred by drone flight.

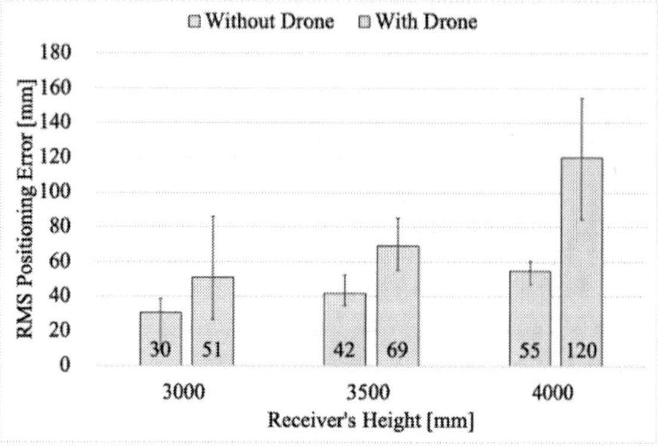

(a) at the center position

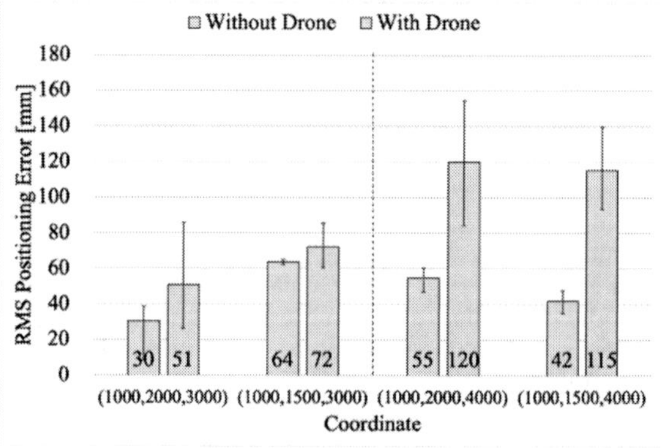

(b) on center position vs.above Tr_2

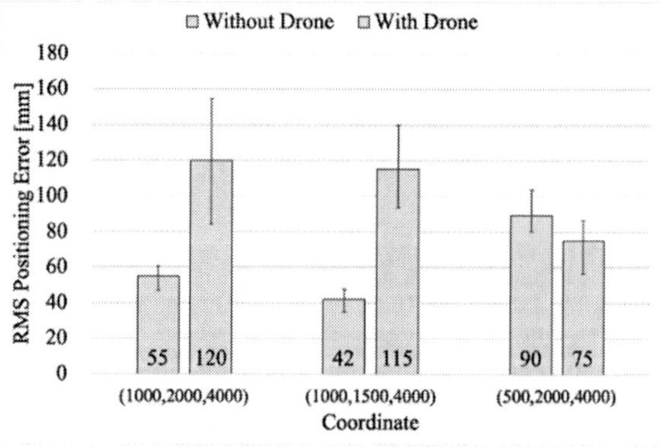

(c) at a height of 4000mm

Figure 11. RMS positioning error.

the installed distances. em_{rms} is defined as

$$em_{\text{rms}} = \sqrt{(dm_i - d_i)^2} \tag{1}$$

where d_i and dm_i are the measured distance and the true distance between a receiver and i-th transmitter, respectively.

Figure 11 shows RMS positioning errors at the same receivers shown in Figure 10 and the maximum and minimum positioning errors as "an expression" of variance. The vertical and horizontal axes of Figure 11 denote the RMS positioning error and the receiver coordinates, respectively. The positioning errors are an average of five trials.

These results indicate that the positioning error increased when the drone is flying because of downwash and flight noise; however, the average errors are less than 15cm. The results of Figures 11(a) and 11(b) confirm that the greater the distance between the transmitter and receiver, the larger the average RMS positioning error and variance when the drone was being flown. Figure 11(c) shows that the most variance is observed at the center of $(1000, 2000, 4000)$.

The results of the experiment indicate that the transmission is accurate enough to measure a drone for a periodic inner wall inspection. We can expect more accurate positioning by compensating for errors caused by the angle of the transmitter and receiver and by the measurement distance [19].

V. CONCLUSIONS

This study proposed a positioning system using SS ultrasonic waves for indoor applications, such as drone communication and wall surface inspection, and evaluated the effects of the system against drone downwash and noise. The proposed SS ultrasonic positioning system transmits and receives SS signals using M-sequence, and the distance is measured using the TOF method. This study mounted small hardware and wideband microphones on the drone. The experimental results for assuming an inner wall inspection by the drone shows that downwash increases the positioning errors, but the errors are less than 15cm. We can expect greater accuracy in the layout of a communication robot because of low DOPs. Therefore, our positioning system using SS ultrasonic waves can be applied for drone application. We will examine the errors in positioning with multiple drones and discuss their errors occurred by flight noise and downwash.

ACKNOWLEDGEMENTS

This work was supported by MEXT KAKENHI Grant Number JP 10609423 and Foster Electric Company, Limited. We would like to thank Uni-edit (https://uni-edit.net/) for editing and proofreading this manuscript.

REFERENCES

[1] L. Apvrille, T. Tanzi, and J.-L. Dugelay, "Autonomous drones for assisting rescue services within the context of natural disasters," in 31st URSI General Assembly and Scientific Symposium, Beijing, China, Aug 2014, pp. 1–4.

[2] A. E. MacDonald, "A global profiling system for improved weather and climate prediction," Bulletin of the American Meteorological Society, vol. 86, no. 12, Dec. 2005, pp. 1747–1764. [Online]. Available: https://doi.org/10.1175/bams-86-12-1747

[3] W. Yoo, E. Yu, and J. Jung, "Drone delivery: Factors affecting the public's attitude and intention to adopt," Telematics and Informatics, vol. 35, no. 6, 2018, pp. 1687 – 1700. [Online]. Available: http://www.sciencedirect.com/science/article/pii/S0736585318300388

[4] U. Nagarajan, G. Kantor, and R. L. Hollis, "Human-robot physical interaction with dynamically stable mobile robots," Proceedings of the 4th ACM/IEEE international conference on Human robot interaction - HRI '09, Mar 2009, p. 281–282. [Online]. Available: http://dx.doi.org/10.1145/1514095.1514176

[5] F. Cunha and K. Youcef-Toumi, "Ultra-wideband radar for robust inspection drone in underground coal mines," in 2018 IEEE International Conference on Robotics and Automation (ICRA), Brisbane,Australia, Mar 2018, pp. 86–92.

[6] A. Suzuki, T. Iyota, Y. Choi, Y. Kubota, K. Watanabe, and A. Yamane, "Measurement accuracy on indoor positioning system using spread spectrum ultrasonic waves," 4th International Conference on Autonomous Robots and Agents, Feb 2009, pp. 294–297.

[7] X. G. et al., "Doppler differential positioning technology using the bds/gps indoor array pseudolite system," Sensors, vol. 19, no. 20, Oct 2019, p. 4580. [Online]. Available: http://dx.doi.org/10.3390/s19204580

[8] D. Kudou, M. Horikawa, T. Furudate, and A. Okamoto, "Indoor positioning method using proximity bluetooth low-energy beacon," in Proceedings of the 17th Asia Pacific Industrial Engineering and Management Systems Conference (APIEMS2016), Taiwan Taipei, Dec 2016.

[9] M. Hazas and A. Hopper, "Broadband ultrasonic location systems for improved indoor positioning," IEEE Transactions on Mobile Computing, vol. 5, no. 5, 2006, pp. 536–547.

[10] L. Segers, A. Braeken, and A. Touhafi, "Optimizations for FPGA-based ultrasound multiple-access spread spectrum ranging," Journal of Sensors, vol. 2020, Jun. 2020, pp. 1–26. [Online]. Available: https://doi.org/10.1155/2020/4697345

[11] A. Suzuki, T. Iyota, and K. Watanabe, "A performance comparison of measurement distance between ook and ss modulation for indoor positioning using ultrasonic transducers," IEEE Sensors, Oct 2011.

[12] ——, "Real-time distance measurement for indoor positioning system using spread spectrum ultrasonic waves," in Ultrasonic Waves. InTech, Mar 2012. [Online]. Available: https://doi.org/10.5772/30215

[13] A. Yamane, I. Taketoshi, C. YoungWoon, K. Yuzuru, and W. Kazuhiro, "A study on propagation characteristics of spread spectrum sound waves using a band-limited ultrasonic transducer," Journal of Robotics and Mechatronics, vol. 16, no. 3, 2004, pp. 333–341.

[14] K. Kumakura, A. Suzuki, , and T. Iyota, "Indoor positioning for moving objects using a hardware device with spread spectrum ultrasonic waves," International Conference on Indoor Positioning and Indoor Navigation (IPIN 2012), Nov 2012, pp. 1–6.

[15] A. Suzuki, K. Kumakura, D. Tomizuka, Y. Hagiwara, Y. Kim, and Y. Choi, "Positioning accuracy on robot self-localization by real-time indoor positioning system with SS ultrasonic waves," Journal of the Korea Society for Power System Engineering, vol. 17, no. 5, Oct. 2013, pp. 100–111. [Online]. Available: https://doi.org/10.9726/kspse.2013.17.5.100

[16] J. Paredes, F. Álvarez, T. Aguilera, and J. Villadangos, "3d indoor positioning of UAVs with spread spectrum ultrasound and time-of-flight cameras," Sensors, vol. 18, no. 2, Dec. 2017, p. 89. [Online]. Available: https://doi.org/10.3390/s18010089

[17] Z. H. et al., "A noise tolerant spread spectrum sound-based local positioning system for operating a quadcopter in a greenhouse," Sensors, vol. 20, no. 7, Apr. 2020, p. 1981. [Online]. Available: https://doi.org/10.3390/s20071981

[18] H. Zhang, G. Chen, Z. Wang, Z. Wang, and L. Sun, "Dense 3d mapping for indoor environment based on feature-point slam method," in Proceedings of the 2020 the 4th International Conference on Innovation in Artificial Intelligence, ser. ICIAI 2020. New York, NY, USA: Association for Computing Machinery, 2020, p. 42–46. [Online]. Available: https://doi.org/10.1145/3390557.3394301

[19] A. Suzuki and T. Iyota, "Angular dependence of transducers for indoor positioning system using ss ultrasonic waves," 2012 International Conference on Indoor Positioning and Indoor Navigation, Nov 2012, pp. 1–6.

Subjective Validity of Bicycle Simulators

Bicycle Simulator Study

Murad Shoman, Hocine Imine
PICS-Lab, COSYS Department, IFSTTAR
Gustave Eiffel University (UGE)
Champs Sur Marne, France
e-mail: murad.shoman@univ-eiffel.fr

Abstract— **In this paper, we briefly present a subjective experimental validation of the bicycle simulator. In the first part, we present the different physical features of the simulator developed at The Perceptions, Interactions, Behaviors & Simulations Lab for road and street users (PICS-L). In the second part, we present the results of an experiment with 10 participants in order to verify the liability of the simulator. For future work, the authors will deploy the simulator in behavioral studies and compare the outputs with experimentation in real environment seeking for the behavioral validity of the simulator.**

Keywords- Bicycle Simulator; Simulation; Subjective Validity.

I. INTRODUCTION

This article is ongoing study started in [1] aiming at improving the dynamics of the bicycle model and simulations to deploy it in future studies on cyclists' behavior and their interaction with different road features in a safe and controllable environment. While our previous work focused on the dynamical modeling of the bicycle simulator, we focus here on the subjective validity of the simulator.

Studying road-user behavior through simulations is a promising tool to address challenges, such as: learning to drive, awareness of risks, road safety, etc. Different simulators have been developed over the past three decades with different focuses and goals [2]-[4]. We can categorize them into: motionless simulators and mobile-based simulators [5]. The first are built around a screen providing visual feedback as in [7], while the latter combines visual information and indices of movements consistent with those of a real vehicle as in [8]-[10]. It has been recognized that often a mobile platform, if well controlled, can significantly improve the realism of the simulation of conduct.

The PICS-L lab at Université Gustave Eiffel designs and develops simulators of conduct to study the behavior of drivers and vulnerable users (cyclists and pedestrians) in different situations, interactions with other users, and information-gathering strategies, in order to better understand the users' immersion in a virtual reality environment [7].

The intermediate complexity and level of realism of the PICS-L bicycle simulator is ensured through several features on the base and background platform, movement control and through numerous sensory information and measurements. These features aim at allowing the cyclists natural behavior, such as: movements, interactions, integrated sound and visual effects similar to the real environment. This equips the simulator for studying cyclists' interactions with other users, road improvements and driving aids. This allows the assessment of the cyclists' subjective risk, perception of and reaction to the infrastructural environment and other road users, as well as anticipation and decision-making. It is noteworthy to state that aspects of balance management and shock situations had been excluded, therefore deferring the study of situations including loss of control and shock, as well as high dynamic demands.

By considering the cyclist as a control system, we could better understand driver behavior and improve the modeling quality of the bicycle and its simulator. In the following, we will focus on the subjective aspects of the bicycle simulator through analyzing different questionnaires. The design of the experiment aims to verify the reliability of the bicycle simulator to validate the model for future experiments. The paper is organized as following. Section 2 is devoted to present the bicycle modeling. In Section 3, experimentation environment and scenarios are detailed, and the results are presented in the Section 4.

II. BICYCLE MODELING

The bicycle dynamic model was created in MATLAB-Simulink, it shows the relations between the different parts of the bicycle model in a graphical format. This allows to graphically trace the various inputs and visualize their relation in MATLAB script format. The model has different sub-layers showing the relative outputs of the different parts. Fig. 1 presents the input /outputs of the Simulink model [1], which was used in the analysis following the simulation.

Eq. (1) was implemented to estimate the friction force:

$$m.\frac{dv}{dt} = F_F + F_a + F_c + F_B + F_g \qquad (1)$$

where m is the total mass of the bicycle-rider system in kg, $\frac{dv}{dt}$ is the longitudinal acceleration as a function of the speed in m/s^2 (the speed was measured using an incremental encoder attached to the rear wheel of the simulator), F_F is the friction force to be calculated, F_c is the force applied by the cyclist on the pedals which is measured using a pedal power meter, F_b is the breaking force and F_g is the gravity force caused by the slopes, F_a is the aerodynamic resistance calculated on the basis of the following equation:

$$F_a = 0.5\, C_{(ax)} \rho\, S\, v_x^2$$

with C_{ax}, the coefficient of aerodynamic resistance given by the bicycle manufacturer, ρ the air density in kg/m^3 and S the frontal surface of the bicycle and the rider body in m^2.

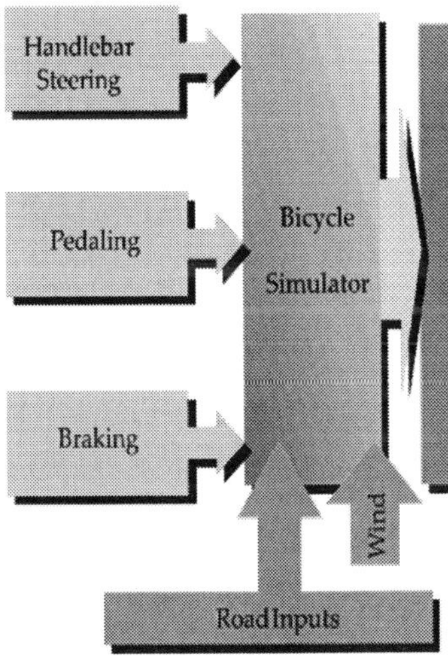

Figure 1. Input/Outputs of the bicycle simulator.

III. Experimental Environment

A. Experimental Setup

The Bicycle Simulator at PICS-L (UGE) with the primary aim to study the bicycle control in virtual environments, was designed by placing a real bicycle on a fixed platform.

The speed of the rear wheel, the angular position of the handlebar, and the gear were measured and logged. A force feedback was applied to the rear wheel using a cylinder in contact with the wheel. The system was composed of a 10N constant part to simulate the dry friction, a variable part to simulate the dynamic friction in proportion to the speed and a variable part proportional to the acceleration to simulate the inertia.

In order to provide more realistic conditions, a fan has been placed in front of the bicycle simulating the wind. The air flow speed was controlled during the simulation as a function of the wheel's speed. The simulated environment consisted of a straight road. Five projectors were fixed in front and beside the cyclist to provide a visual cue with 225° of horizontal and 55° of vertical field of view. A rear screen was placed behind the bicycle to allow a rear-view if the cyclist needs to turn around. The audio cue is simulated by software with the use of an audio system consisting of four speakers and a subwoofer reproducing the sound of the rolling tires (using speed), wind (using speed) and traffic. The simulator is equipped with gear, clutch and braking pedals. Fig. 2 shows the different features of the bicycle simulator.

B. Participants

Ten subjects (6 male; mean age=28.17, SD=3.76 and 4 females; mean age=25.25, SD=2.06) participated in this experiment. All had normal or corrected-to-normal vision. The mean cycling experience of the participants was 12.9 years. The average number of cycling kilometers per month was 62.

Figure 2. A subject during the experiments.

C. Scenario

The experiment took place in a simulated urban environment. The road consisted of two straight sections: the first was a bicycle-bus shared lane, the latter a separate bicycle lane. Traffic was generated in the same and opposite direction of the cyclist and buses were passing the cyclist from time to time. The participants were asked to take a pre-ride in order to familiarize themselves with the simulator. They were asked to maneuver with the simulator and use the different features, such as: handlebar, pedals, gear and brakes. Following the pre-ride, the participants were asked to go for a ride on the bicycle simulator along the virtual street.

The duration of the experiment was around 10 minutes, which we considered long enough to test all the features of the simulator, collect enough date for post-analysis and not exhausting for the participants. The results of the simulation are used to validate the theoretical and the physical model of the bicycle simulator.

At the end of the experiment, the participants filled three questionnaires. The first questionnaire consisted of general information and cycling experience of the participants in real life and using the simulator, followed by the Simulator Sickness Questionnaire (SSQ) [12]; the participants evaluated their experience, through 16 questions, indicating on a scale of four steps (None, Slight, Moderate and Severe) the occurrence of different symptoms during the experiment. The third questionnaire, NASA Task Load Index (TLX) [13], was aimed at evaluating the overall workload of the cycling task, and the importance of each of the 6 work-load-factors under investigation. The questionnaires were available both in English and French as some participants speak only French.

D. *Simulation outputs*

Fig. 3 shows the speed profile of the bicycle simulator for one of the subjects measured through the incremental encoder attached to the rear wheel of the simulator. We could notice the acceleration and breaking phases at the beginning and end of the experiment.

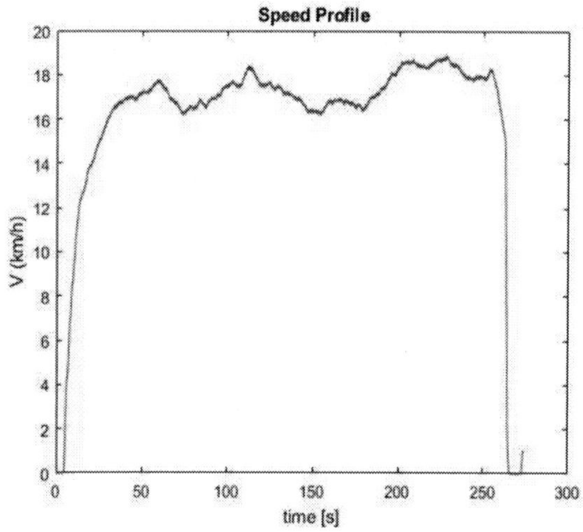

Figure 3. Speed profile of one of the participants in km/h.

The steering angle of the bicycle simulator handlebar, shown in Fig. 4, is measured using an incremental encoder. It is noticed that the steering angle ranges between +\- 4, which is relatively small. This could be explained due to the straightness of the virtual road as noticed in the trajectories and global position (Fig. 5 and 6).

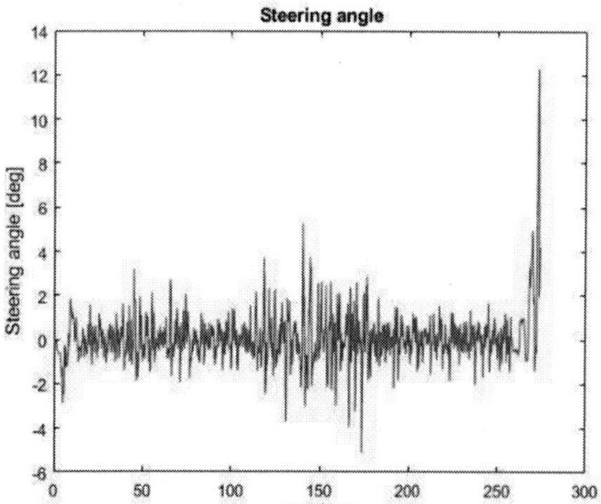

Figure 4. Steering angle of the simulator handlebar.

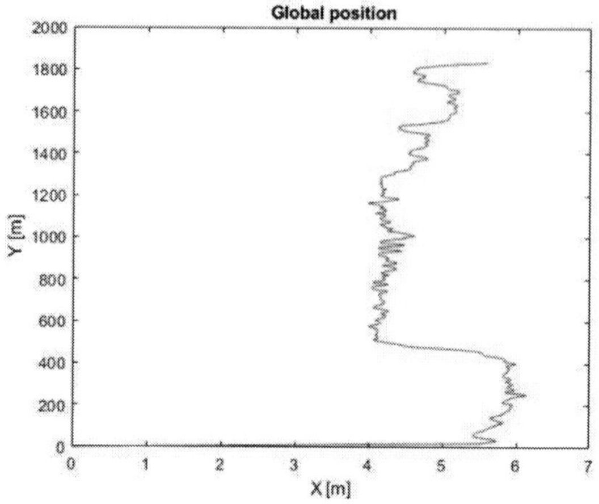

Figure 5. Global position of the simulator extracted from the virtual environment.

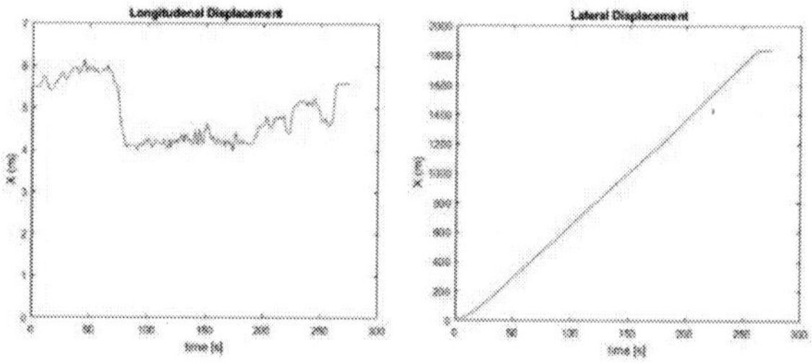

(a) (b)

Figure 6. (a) The longitudinal and (b) the lateral trajectory of the simulator extracted from in reference to the virtual environment coordinates.

The estimated friction force calculated using Eq. (1) is shown in Fig. 7.

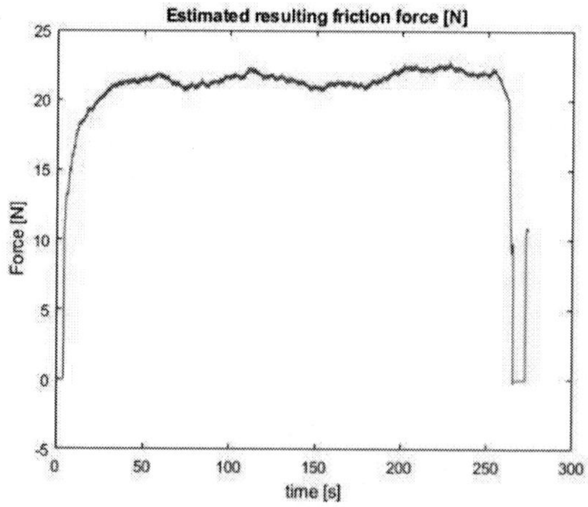

Figure 7. Estimated friction force (N).

IV. SUBJETIVE VALIDITY RESULTS

A. Cycling Experience Questionaire

The results of the first questionnaire are summarized in Appendix 1. As noticed in the last column, the average evaluation of the similarity to cycling in real environment ranges between 4 and 8 on scale of 10 (mean= 6.1, SD= 1,6). Assessing the realism of the simulator, participants mentioned the good design of the virtual road, pedaling, traffic and other sensory ques, such as the wind and the sound of the passing traffic. As suggestions for improvement, most participants agreed they expect higher speed compared to the cycling effort.

B. Simulator Sickness Questionaire (SSQ)

The analysis of the simulator sickness questionnaire listed in Table 1 shows that the average total severity of all participants is around 32.5. By comparing this number to the possible scores listed in Table 2, we could see that the total severity of the simulator is slight (less than 78.5). It is also noticed that participants wearing lenses experienced the highest total severity (participant #6 has 115.9 and participant #10 has 71). The affected participants showed high disorientation symptoms.

TABLE I. ANALYSIS OF THE SIMULATOR SICKNESS QUESTIONNAIRE (SSQ).

Participant number	Total severity	Oculomotor	Nausea	Disorientation
1	7.48	0	7.58	13.92
2	33.66	9.54	37.9	41.76
3	7.48	9.54	7.58	0
4	0	0	0	0
5	14.96	9.54	22.74	0
6	115.94	114.48	37.9	194.88
7	63.58	38.16	53.06	83.52
8	11.22	9.54	7.58	13.92
9	0	0	0	0
10	71.06	38.16	68.22	83.52
mean	**32.538**	**22.896**	**24.256**	**43.152**
SD	36.80	35.18	23.92	62.75
Min	0	0	0	0
Max	115.94	114.48	68.22	194.88

TABLE II. POSSIBLE SCORE RESULTS OF SSQ.

	Nausea	Oculomotor	Disorientation	Total Severity
none	0	0	0	0
slight	66.8	53.1	97.4	78.5
moderate	133.6	106.1	194.9	157.1
severe	200.3	159.2	292.3	235.6

The average exposition to different symptoms during and after riding the bicycle simulator are reported in the radar chart view (see Fig. 8)

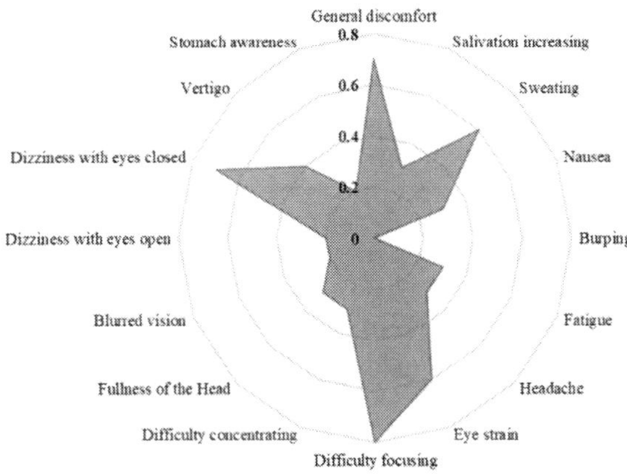

Figure 8. Mean scores observed in each item of the exposure of simulator sickness questionnaire.

C. NASA Task Load Index (TLX)

The Nasa Task Load index is used to collect subjective workload assessments for different simulators [13]. Table 3 shows the weighted ratings of the NASA TLX. The first column shows the scales under assessment, the second represents the average weight of each scale according to the personal opinion of the participant. The scorer chooses different factors on an evaluation cards according to its importance, while the weight of each factor is the number of times it was circled. The third column is the average raw rating taken from the TLX questionnaire, where the participants evaluated each factor on a scale of 100. The last column represents the adjusted weighting, which is the multiplication of the weight and raw rating of each factor. It is noticed that the physical demand was highly weighted affecting the overall work load (87.4 on a scale of 100), while the raw rating shows a moderate overall workload.

TABLE III. WEIGHTED RATING OF TLX.

Scale title	Weight	Raw Rating	Adjusted rating
Mental Demand	3	27.78	83.33
Physical Demand	4	47.78	191.11
Temporal Demand	2	31.11	62.22
Performance	1	42.22	42.22
Effort	3	34.44	103.33
Frustration	2	21.11	42.22
Overall workload		**34.07**	**87.41**

Fig. 9 shows the weighted average of each work load factor; the width of each column represents the weight of each factor. We notice the performance was weighted the least, this could be explained because the required task was simple and easy to accomplish, so the participant chose not to give it a high raring.

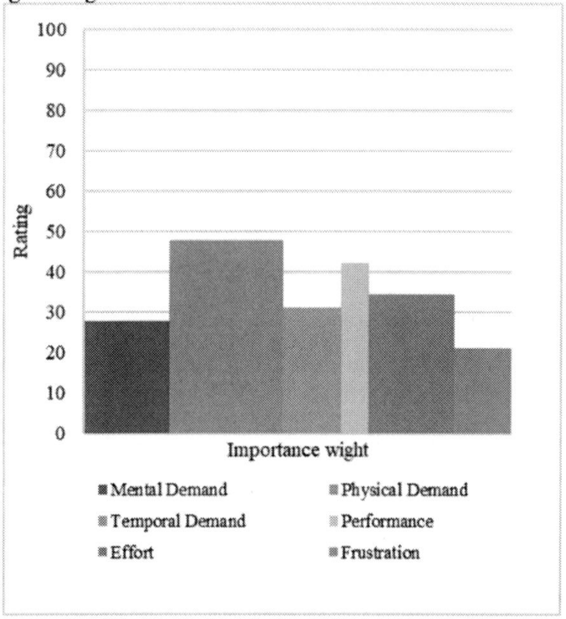

Figure 9. Graphic representation of the composition of a weighted workload score.

V. CONCLUSION AND FUTURE WORK

The bicycle simulator enables us to put cyclists in a riding situation and accurately measure their effective behavior, while controlling the variables at play and avoiding the risks associated with a real environment.

After analyzing the different questionnaires, it is possible to verify the reliability and to subjectively validate the simulator. The results show low simulator sickness and relatively high workload, which could be explained by the effort done by the cyclist.

Further development was applied to the simulator including mathematical model improvements, development of the virtual environment and installment of new devices to simulate the interaction between the infrastructure and the bicycle. An additional experiment, including 36 subjects, was conducted in order to validate the new model physically and subjectively. The results of the new experiment will be published in an upcoming paper.

ACKNOWLEDGMENT

This works is funded by Marie Skłodowska-Curie actions (H2020 MGA MSCA-ITN) within the SAFERUP project (grant agreement number 765057). The authors gratefully acknowledge their contributions.

REFERENCES

[1] M. Shoman and H. Imine, "Modeling and simulation of bicycle dynamics ", In TRA 2020, Transportation Research Arena, TRA, 2020, doi:10.5281/zenodo.3775338.

[2] N. Ghasemi et al., "Longitudinalmotion cueing effects on driver behaviour: a driving simulator study", Advances intransportation studies, vol. 49, pp. 91-102, 2019.

[3] A. T. Vu, M. T. Nguyen, D. V. M. Nguyen, and V. Hung Khuat, "Investigating the effect of blood alcohol concentration on motorcyclists riding perfor-mance using an advanced motorcycle simulator", Transportation Research Part F: Traffic Psychology and Behaviour, vol. 73, pp. 1–14, 2020.

[4] J. Boril, M. Jirgl, and R. Jalovecky, "Use of flight simulators in analyzing pilot behavior", In IFIP International Conference on Artificial Intelligence Applicationsand Innovations, pp. 255–263, Springer, 2016.

[5] A. Bucchi, C. Sangiorgi, and V. Vignali, " Traffic psychology and driver behavior", Procedia-social and behavioral sciences, vol. 53, pp. 972–979, 2012.

[6] L. Nehaoua, "Conception et réalisation d'une plateforme mécatronique dédiée à la simulation de conduite des véhicules deux-roues motorisés", PhD diss., 2008.

[7] S. V. Babu et al., "An immersive virtual peer for studying social influences on child cyclists' road-crossing behavior", IEEE transactions on visualization and computer graphics, vol. 17(1), pp. 14–25, 2011.

[8] D. S. Kwon et al., "Kaist interactive bicycle racing simulator the 2nd version with advanced features", IEEE/RSJ International Conference on Intelligent Robots and Systems, 2002.

[9] Q. He, X. Fan, and D. Ma, "Full bicycle dynamic model for interactive bicycle simulator", Journal of Computing and Information Science in Engineering, vol. 5(4), pp. 373–380, 2005.

[10] C. K. Chen, F. J. Chen, J. T. Huang, and C. J. Huang, "Study of interactive bike simulator in application of virtual reality", Journal of the Chinese society of mechanical engineers, vol. 28(6),pp. 633–640, 2007.

[11] S. Caro, N. Chaurand, N. T. Dang, and F. Vienna, "Design of a traffic simulator for the study of the behavior of cyclists", Journal of Transport - Dissemination (JTD) of the Scientific and Technical Network, 2013.

[12] R. S. Kennedy, N. E. Lane, K. S. Berbaum, and M. G. Lilienthal, "Simulator sickness questionnaire: An enhanced method for quantifying simulator sickness",The international journal of aviation psychology, vol. 3(3), pp. 203–220, 1993.

[13] S. G. Hart, "Nasa task load index (TLX) ", vol. 1.0; computerized version, 1986.

APPENDIX I : PARTICIPANTS RESPONSES TO CYCLING EXPERIENCE QUESTIONNAIRE.

Participant Number	**Gender**	**Vision**	**Age**	**Avg. cycling per month (km)**	**Cycling experience (years)**	**Realism of simulator* (Scale of 10)**
1	M	Normal	26	20	20	4
2	F	Glasses	27	3	3	4
3	M	Normal	27	21	21	8
4	M	Normal	30	6	6	6
5	M	Normal	35	7	7	7
6	F	Lenses	23	10	10	6
7	F	Normal	27	20	20	7
8	M	Normal	26	16	16	8
9	M	Normal	25	10	10	7
10	F	Lenses	24	16	16	4